Leif Glaser

Magnetic Properties of Deposited CoPt Clusters and Nanoparticles

Leif Glaser

Magnetic Properties of Deposited CoPt Clusters and Nanoparticles

XAS and XMCD studies at the Cobalt L-edge of in situ prepared size selected CoPt clusters and wetchemically synthesized CoPt nanoparticles

Südwestdeutscher Verlag für Hochschulschriften

Impressum/Imprint (nur für Deutschland/only for Germany)
Bibliografische Information der Deutschen Nationalbibliothek: Die Deutsche Nationalbibliothek verzeichnet diese Publikation in der Deutschen Nationalbibliografie; detaillierte bibliografische Daten sind im Internet über http://dnb.d-nb.de abrufbar.
Alle in diesem Buch genannten Marken und Produktnamen unterliegen warenzeichen-, marken- oder patentrechtlichem Schutz bzw. sind Warenzeichen oder eingetragene Warenzeichen der jeweiligen Inhaber. Die Wiedergabe von Marken, Produktnamen, Gebrauchsnamen, Handelsnamen, Warenbezeichnungen u.s.w. in diesem Werk berechtigt auch ohne besondere Kennzeichnung nicht zu der Annahme, dass solche Namen im Sinne der Warenzeichen- und Markenschutzgesetzgebung als frei zu betrachten wären und daher von jedermann benutzt werden dürften.

Verlag: Südwestdeutscher Verlag für Hochschulschriften GmbH & Co. KG
Dudweiler Landstr. 99, 66123 Saarbrücken, Deutschland
Telefon +49 681 37 20 271-1, Telefax +49 681 37 20 271-0
Email: info@svh-verlag.de

Approved by: Hamburg, Universität, Diss., 2009

Herstellung in Deutschland:
Schaltungsdienst Lange o.H.G., Berlin
Books on Demand GmbH, Norderstedt
Reha GmbH, Saarbrücken
Amazon Distribution GmbH, Leipzig
ISBN: 978-3-8381-1726-3

Imprint (only for USA, GB)
Bibliographic information published by the Deutsche Nationalbibliothek: The Deutsche Nationalbibliothek lists this publication in the Deutsche Nationalbibliografie; detailed bibliographic data are available in the Internet at http://dnb.d-nb.de.
Any brand names and product names mentioned in this book are subject to trademark, brand or patent protection and are trademarks or registered trademarks of their respective holders. The use of brand names, product names, common names, trade names, product descriptions etc. even without a particular marking in this works is in no way to be construed to mean that such names may be regarded as unrestricted in respect of trademark and brand protection legislation and could thus be used by anyone.

Publisher: Südwestdeutscher Verlag für Hochschulschriften GmbH & Co. KG
Dudweiler Landstr. 99, 66123 Saarbrücken, Germany
Phone +49 681 37 20 271-1, Fax +49 681 37 20 271-0
Email: info@svh-verlag.de

Printed in the U.S.A.
Printed in the U.K. by (see last page)
ISBN: 978-3-8381-1726-3

Copyright © 2010 by the author and Südwestdeutscher Verlag für Hochschulschriften GmbH & Co. KG and licensors
All rights reserved. Saarbrücken 2010

Contents

1 **Introduction**   3

2 **Fundamentals**   16
    2.1 X-Ray Absorption Spectroscopy 16
        2.1.1 X-Ray Magnetic Circular Dichroism 25
        2.1.2 Thermal Desorption Spectroscopy 32
        2.1.3 Soft Landing 34
        2.1.4 STM 35
    2.2 Material properties 37
        2.2.1 $3d$ transition metals 37
        2.2.2 Cobalt-Platinum alloy 43
        2.2.3 Magnetic thin films 47

3 **Experimental Setup**   55
    3.1 Introducing wet chemical CoPt nanoparticles 56
        3.1.1 Sample preparation and characterization . . . 58
    3.2 Mass selected Co_nPt_m clusters 65
        3.2.1 Sputter chamber and mass selection 65
        3.2.2 Sectroscopy chamber 71
        3.2.3 In situ sample preparation 75
    3.3 Data acquisition 78

	3.3.1	Experiments at a storage ring	79
	3.3.2	Tey measurement	82
	3.3.3	Tey background treatment	85

4 Wet Chemical Nanoparticles — **90**
- 4.1 Oxidation effects 90
- 4.2 Branching ratio 107
- 4.3 Ratio of orbital to spin magnetic moment 109
- 4.4 Orbital and spin magnetic moment 114
- 4.5 Discussion 118

5 Mass Selected Clusters — **127**
- 5.1 Iron substrate 128
- 5.2 XAS whiteline spectra 138
- 5.3 XAS dichroism spectra 147

6 Summary — **161**
- 6.1 Conclusion 161
- 6.2 Outlook 165

A Mass spectra — **I**

B Abbreviations — **XII**

List of figures — **XVIII**

List of tables — **XIX**

Bibiliography — **XIX**

Chapter 1

Introduction

Since their discovery magnetic systems have always been of high interest to humans. Around 1100 BC the Chinese used the magnetism of magnetite (Fe_3O_4), the earliest signs of it's use in Europe is documented by Thales (624 - 546 BC). It took two millennia for the first revolutionary application to evolve: The compass was documented as wet compass in China in the 11th century AD and as dry compass in Europe in the 13th century AD. Only further 500 years later the experiments of Oerstedt, Ampère, Biot and Savat in 1820 AD and Faraday in 1821 started a new era of magnetism: Electromagnetism and its applications, as for instance the generator, the transformer and the loudspeaker. As a particular application of electromagnetism conventional electromagnetic storage media technology was established over the years. In tapes, disks and hard drives coil driven electromagnets where used to read and write the information. Theses rather large coils limited the minimization process and hence the maximal storage capacity. The discovery of the Giant Magneto Resistance [18, 45] (GMR) by P.Grünberg and A.Fert in 1988 opened the door to the minimization of the read head[1] and thus higher

---
1. European Patent Number 0346817

storage density. In 1997 the first commercial hard drive with the GMR-technology was presented by IBM. Reduction to 50 nm sized magnetic storage cells and thus storage density's of 1 Terabit per square inch is proposed by Hitachi for 2010.

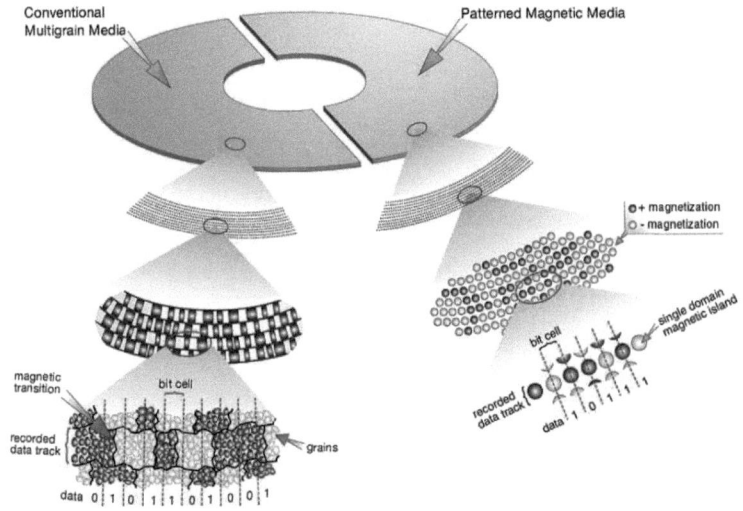

Figure 1.1: The conventional storage media change from multigrain to patterned media (figure from reference [59]). This implies the necessity to find useful small particles to synthesize new high density storage discs.

Typical hard drives are build up out of granular media [2] and is successively replaced by patterned media (figure 1.1). The rather huge size distribution of the grains make it difficult to decrease the size of one bit, still securing a proper identifiable signal. Hence the trends goes to patterned material that is structured in very similar sized units. For these units one could potentially use magnetic nanopar-

---
2. A typical material in the anti-ferromagnetic storage technology, that is still used is $Co_{70}Cr_{18}Pt_{12}$

ticles that distribute nice in a close package on a surface, either as a thin film by self organization or in for example a polymer matrix.

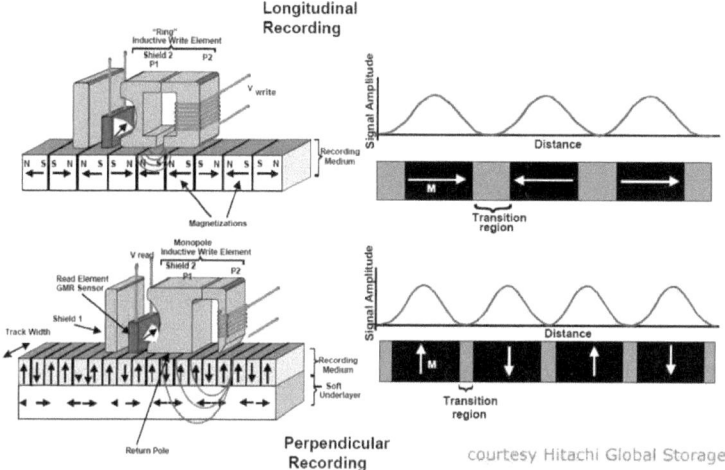

Figure 1.2: Former longitudinal (in plane magnetized) storage must be replaced by perpendicular storage, since the increasing size reduction can be accomplished better with the perpendicular recording (figure from reference [59])

Additionally to the decreasing size of the grains the recording technology from formally in plane magnetized bits, is changed to perpendicular magnetized bits. As shown in figure 1.2 the longitudinal storage technology consumes more lateral space, due to a larger required transition region and typically slightly larger required "bit" sizes. The future of magnetic storage application requires small magnetic particles with narrow size distribution and perpendicular magnetization of the storage material.

To create small magnetic units there can be two approaches: Top-down[3] or bottom-up[4]. Starting with the first approach, by cutting larger objects into small units the standard lithography techniques is immersion lithography using deep ultra violet (DUV) light (193 nm/6.4 eV) or in the future extreme ultra violet (EUV) light (13.5 nm/92 eV) and Electron beam lithography. Electron beam lithography has a proximity limit due to scattered electrons of about 20 nm lateral resolution and transfers charges to the object, which may cause difficulties with insulating material or semi conductors. Using light for cutting is mainly limited by the light sources available and by the amount of energy the target material can sustain, the cleanliness of wafers and masks, as well as the vibration stability of the lithographic device. IBM has developed a test Immersion lithography device with DUV reaching lateral resolution of 30 nm [65]. INTEL plans a 22 nm resolving factory with EUV for 2011, while SEMATECH (SEmiconductor MAnufacturing TECHnology) has demonstrated a 22nm half pitch [120][5] in 2008, where 15 mJ/cm^2 was applied. While the lateral resolution was good, the line width roughness (LWR) was 5-6 nm, which is far above the usual 3% expected for industrial applications [55]. The International Technology Roadmap for Semiconductors (ITRS) publishes expectations [71] that lithographic systems may reach 22 nm industrial applicability in 2016 and that lower resolutions will most likely need directed self assembly, with the goal of 11 nm resolution in 2022.

Building larger scale objects from small subunits the assembling may

---

3. take a lager piece and cut it into appropriate sized subunits
4. start with small particles and assembles a larger scale system
5. typical test object in lithography, where grooves are cut into an object and both grooves and bars of the cut objects have the same width.

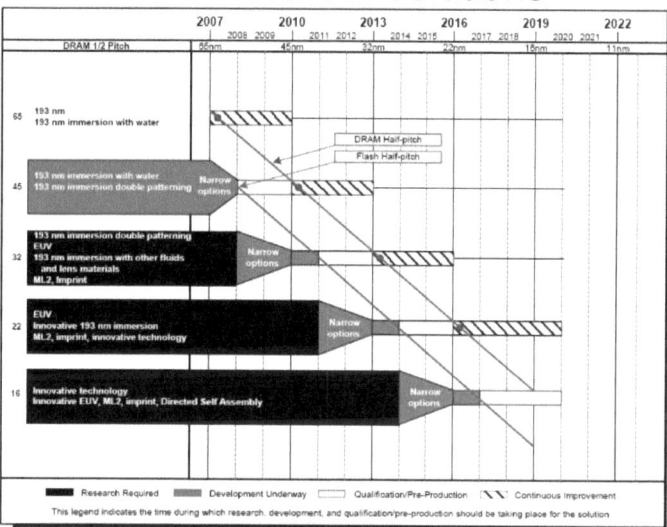

Figure 1.3: ITRS Roadmap 2008 (figure from reference [71]) shows the estimation on development of the top down size regime. The techniques used today allow the structuring in a size regime of about 50 nm. In the next 15-20 years this size regime may go down to about 10-15 nm structures. This is believed to be the lower limed with lithographic techniques. for smaller structures different technical approaches as for example self assembling nanoparticles or the help of virus based construction will be needed.

still proof difficult. For test cases a proof of principle structure can be assembled and investigated by means of STM (scanning tunneling microscope). For large scale application spontaneous or directed self assembling of the systems is unavoidable. There have been multiple reports of self assembling [124, 133, 148] and recently some first results have been achieved using the common *Tobaccomosaicvirus*, which is 18 nm in diameter and 300 nm long [87, 91, 142], to produce and deposit metallic nanotubes.

Wet chemical or biological synthesis of metallic nanoparticles has the advantage of large scale applicability, but usually combined with the effect that some organic material is attached to the particles. Whether these ligands can be removed and whether the removal or the remain of those ligands affects the desired particle properties has to be investigated in each case. Organic spacing layers may even be desired to separate two adjacent magnetic units.

In a top down approach cutting well known material into smaller pieces conventionally the size effects upon the material are gradual while the pieces become smaller with improvements in cutting technology. Assuming nearly complete magnetic orientation of a storage cell, when keeping shape and material constant, but scaling the cell size, it is obvious that the smaller the magnetic cell, the smaller is the total magnetic energy stored in the system. In case of small storage unit sizes, one can reach the superparamagnetic regime, where all atoms of one particle are magnetically ordered in one direction (typically particles of a few nm Diameter are superparamagnetic). The total energy that fixes a magnetization in a superparamagnetic particle used as a bit, is therefore the average magnetization anisotropy of one atom in the particle times the number of atom of one particle. When this energy is in the range of the thermal energy available the magnetization flips randomly and no storing of information is possible. The temperature can therefore either be lowered far enough (below the so called blocking temperature) that the thermal activation energy is no more sufficient to change the magnetization of the particle. A mandatory cooling system to preserve data on a mass produced storage device is not the most desired option. Storage devices should typically be used at room temperature and hence the thermal energy is fixed, therefore one needs to try and raise the Magnetic Anisotropy Energy (MAE) of the particle until the Energy needed to flip the magnetization of the particle is above the thermal activation energy available at room temperature (25 meV/atom). An extremely high value measured so far for a single Co atom on a Pt surface was 9 meV [50]. One needs at least several 100 atoms to achieve magnetically long term stable structures. The typical goal is to ensure a lifetime of recorded information of at least 10 years for data stored on a standard hard drive.

Most of the magnetic moment of a typical magnetic particle is stored in the spin magnetic moments of the atoms involved. A significantly smaller portion is stored in orbital magnetic moments, which are aligned by coupling to the crystal field (geometry). The spin moments couple by spin-orbit interaction to the orbital moment and thus indirectly to the crystal field.

Trying to minimize the size of one bit, it is obviously important to increase the average MAE per volume of the bit to keep the total MAE constant or increase it. Materials that have high magnetization with strong alignment to the crystal lattice are therefore prime candidates to investigate. $3d$ metals show strong magnetism, while their spin orbit interaction is rather small, while heavier metals as $4d$ and $5d$ elements have less magnetism, but higher spin orbit coupling. Materials that fulfill both, high magnetization and high magnetic anisotropy, are for example $3d$-$5d$-alloys as Co_xPt_{100-x}, where the high magnetic moment of the $3d$ metal is coupled by $3d$-$5d$ hybridization to the platinum and thus to the strong spin orbit coupling of the $5d$ element and further to the crystal field. There are numerous candidates for these alloys. The most promising and much investigated for over 20 years were FePt and CoPt alloys. Both alloys show in all different stoichiometric configurations high magnetic anisotropy and high coercivity.

Most investigations were done using bulk or thin film materials, if the material is additionally covered with some organic shell or it is highly structured (as into nanoparticles) it may change the total magnetization of the system, but likewise change it's chemical reactivity. Hence it is very important to investigate the effect that these changes may have on the material and what happens, when these material are produced in conditions close to mass-production. On the other hand, one might ask, what influence the supporting material may have on the very small particles, because as any device that consists out of nano-scaled units has to be macroscopically stabilized.

In case of oxidation effects of particles with shrinking sizes, one could cover the surfaces with for example gold (patented by US Pat.No. 7226636 by the Los Alamos National Laboratory). The coverage by chemically inert materials could open other fields of application for small magnetic particles, as in medical applications.

For medical applications for instance Co as an element may not be used due to it's toxic effects on biological systems, but where small gold nanoparticles were already successfully used and gold covered particles could be applied.

In dealing with new aspects of materials there is always the fundamental question of understanding the physical interaction. In case of magnetism Hund's rule predicts a maximized orbital magnetic moment for atoms, while in transition metal bulk material the orbital moment is almost completely quenched due to band structure formation. The interesting question of how the properties change between atom and bulk (figure 1.4) and how to describe this properly in models is still not answered. For larger systems (above 10 nm) it is commonly accepted that the particles properties change rather smoothly. In this size regime the so called liquid drop model is used (figure 1.4), which is a classical electrostatic model, describing the particle as uniform sphere, while atomic positions are ignored. Particles of 10 nm and more consist of well above 50 thousand atoms and are considered fairly large (from a cluster physicist point of view). In some cases (especially if the valence electrons of the material are very weakly bond) a very similar model, the jellium model can describe some properties even down to a few atom sized clusters. The jellium model uses a uniformly positively charged sphere that is filled with an (delocalized) electron gas. This model predicts magic cluster sizes for electronic or geometric shell closures. Most systems are less well behaved and the regime of the quantum size effects is rather uncharted territory. The best way to investigate such a systems and benchmark theoretical models is to prepare it as precise as possible and then change if possible only one parameter at a time. For comparison with theory it is of course important that the systems can be theoretically described and calculated. In the case of small systems quite often the electron correlation effects play a major role and therefore more electrons are to be taken to account in the calculations. If effects of non collinear coupling of spins are included in the theoretical calculations the required computational power rises even more. In the case of experimental and theoretical investigations of small mass selected chromium clusters [89] it has been proven to be impossible with the computers at this point of time to fully relativistically calculate Cr-clusters of sizes as "small" as 10 atoms in contact to a surface, including electron correlation effects. Similar problems occur if more than one type of atoms is among the cluster material. It is very demanding for the theoretical physicists to describe cluster material like Co_xPt_{100-x} including electron correlation effects. This makes it even more important for experimentalist to deliver measurements as a basis to judge the first results of theoretical models and calculations.

It is well known, that free cobalt atoms have a magnetic moment of $3\mu_B$ and that only 1.75 μ_B [16] remains in cobalt bulk material, but much of how this change takes place and what happens if the atoms, cluster or nanoparticles (figure 1.4) are deposited on surfaces must still be investigated. Above that the effects of the stoichiometric composition of an alloy as Co_nPt_m will most likely add some new effects that need to be understood, when trying to design new material for any application.

This work presents a part of the world wide process of gathering experimental understanding of size and compositional property changes of small magnetic clusters and particles. On the path to-

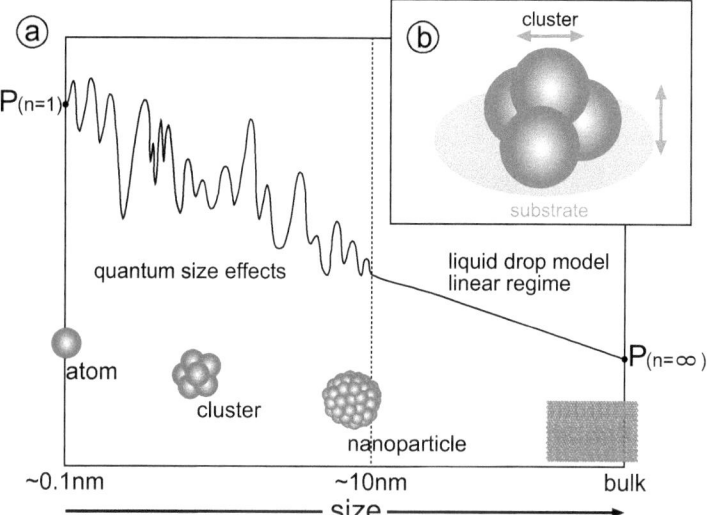

Figure 1.4: (a)Size effects: Changing a particle size, when investigating one property of a material, it starts at small sizes in the regime dominated by cluster size effects and ends at bulk material. For larger particles the liquid drop model can most often be used, where the particles properties change smoothly. Typically the size of 10 nm which is several 10 thousands of atoms is considered to be the turning point for the change from quantum size effect regime to smooth liquid drop model regime, but where and how this takes place and what happens below the 10 nm threshold is mostly unknown. (b) Additionally to changes from within the particle due to it's varying size, in real systems particles have always contact to some supporting material, therefore the influence of these support materials upon the supported particle is very important.

wards the ultimate goal of tailoring magnetic materials for future applications, the understanding of fundamental interactions are an essential first step. Experimental investigation of single parameter induced changes of the magnetic properties in well characterized cluster systems may help to improve and benchmark theory. Still much effort has to be devoted to this course to establish a broad basis set of experimental data. Apart from the basic research detailed investigation of large scale applicable material upon any influences linked to it's magnetic properties may help optimize production processes and to develop new approaches for future material synthesis. It is of great importance to find out, what the limits of a certain production process are. If theoretically well functional materials can not be produced as calculated, either the production process must be adjusted or the material must be designed according to the limitations of the synthesis process.

Following the two major aspects of gathering data for basic research and investigating actually large scale producable material. Two different sample types were investigated , small clusters of 1 to 4 atoms per cluster and small nanoparticles of 3.7 nm to 8.4 nm in diameter. The small clusters are an ideal model system that are not too large to be calculated by theory and simple enough to be experimentally prepared in well defined samples. The clusters preparation method requires ultra high vacuum (UHV) conditions and in situ preparation, can thus not be expanded to produce macroscopic amounts of sample material. The nanoparticles were produced with wet chemical methods and thus represent a possible class of technical applicable substances. Wet chemical methods allow macroscopic amounts of sample material to be produced and usually require less experimental expenditure than physical UHV based sample preparation methods, but carry the drawback that not all parameters can be controlled separately and most often not even all parameters are known. Above that the nanoparticles are too large to be calculated in an exact model by theory, but investigating those systems is essentially to learn about applicability of any new sample material. As primary investigation tool x-ray magnetic circular dichroism (XMCD) was chosen, since it allows element selective investigation of a mixed target systems and allows to measure spin and orbital moments separately.

In chapter 2 all experimental methods used will be introduced, followed by an introduction of the materials used in the experiments. The different samples, their preparation and experimental setup at the storage ring will be discussed in chapter 3. Wet chemically prepared CoPt Nanoparticles will be presented and discussed in chapter 4, where the aspects concerning aging played an important role. The following chapter 5 deals with the results of the mass selected CoPt clusters, where aging was excluded by fresh in situ preparation of all samples, but other aspects due to precise control ability of the preparation parameters could be investigated. The work will be concluded by a short summary and outlook in chapter 6.

Chapter 2

Fundamentals

This chapter will give an overview of the experimental techniques used for the experiments performed in this work, followed by a short introduction of the materials used in the experiments and some of their relevant properties.

2.1 X-Ray Absorption Spectroscopy

X-ray absorption spectroscopy (XAS) is an element specific spectroscopic method for the analysis of the electronic structure of a material. A target is illuminated with monochromatic light of a x-ray source. Light will be absorbed by the target to excite electrons into unoccupied bound states of the material (figure 2.4a) or to free continuum states (figure 2.4b). Choosing the energy used for this technique in the order of the binding energies of core electron of atoms, mainly those are excited. Core electrons are localized and have element specific binding energies, thus making XAS element specific (table 2.1).

When exciting electrons in a single photon absorption process, the dipole selection rule will be followed $\Delta l = \pm 1$ and $\Delta s = 0$ and additionally for polarized light $\Delta m = +1/0/-1$ *(right circular / linear / left circular)*. The orbital moment of initial and final state differ by one and there is no spin flip in a dipole transition. The magnetic quantum number m may stay or change by one, depending on the polarization of the light. By tuning the energy of the light, while measuring the absorption, a signal proportional to the energy dependent absorption cross section $\sigma(E)$ of the material can be measured. The absorption cross section is defined as the ratio of exited electrons per unit time $T_{i \to f}$ and the photon flux I_{ph}:

$$\sigma(E) = \frac{T_{i \to f}}{I_{ph}} \qquad (2.1)$$

In a single particle picture the number of excited electrons per unit time is given with time dependent perturbation theory and known as Fermi's Golden rule:

$$T_{i \to f} = \frac{2\pi}{\hbar} \langle f|H'|i \rangle \cdot \rho(E) \cdot \delta(E_f - E_i - E_{ph}) \qquad (2.2)$$

In which the energy of the light $E_{ph} = \hbar\omega$, the initial state $|i\rangle$ and final state $|f\rangle$, the energy difference of final and initial state $E_f - E_i$, $\rho(E)$ the energy density of the final state and H' describes the perturbation of the Hamilton operator due to the electromagnetic field

element	ground state	L_3-edge (eV)	L_2-edge (eV)	lattice	lattice constant (Å)
Fe	$[Ar]3d^6 4s^2$	706.8	719.9	bcc	2.870
Co	$[Ar]3d^7 4s^2$	778.1	793.2	hcp	2.510
Ni	$[Ar]3d^8 4s^2$	852.7	870.0	fcc	3.520
Cu	$[Ar]3d^{10} 4s^1$	932.7	952.3	fcc	3.610

Table 2.1: Some atomic and bulk properties of Fe, Co, Ni and Cu

of the photons. Using the dipole approximation (assumed constant electromagnetic field in the interaction region $k \cdot r \ll 1$) and taking only single photon absorption processes into account H' can be simplified to to the product of the polarization vector of the light and the position operator: $H' = \epsilon \cdot r$

In a simple picture the absorbing atom emits the photoelectron in form of a spherical wave, that will be backscattered by all surrounding atoms. These backscattered waves interfere with the initial wave, constructively or destructively, depending on the phase shift of the backscattered waves. If the structure has some long range order there will be some periodical dependence of the intensity of the outgoing wave, depending on the structure of absorber and on the energy of the incoming light. The structure influences mainly the angular dependence of the intensity of the escaping electrons at constant incoming photon energy, while changes of energy of the incoming light will change the energy of the outgoing electrons and thus their wavelength and hence the interference pattern. The observation of these periodic changes of the absorption coefficient due to these backscattering events is typically possible up to several 100 eV above the absorption edge and is called extended x-ray absorption fine structure (EXAFS). This can be seen schematically in figure 2.1.

The EXAFS oscillations are used to investigate the symmetry of the local environment of the absorbing atoms, as well as it's density of unoccupied states, next neighbor distance and coordination number. More detailed information may be found in the literature [58, 132]. XAS can best be measured in transmission for samples with a thickness in the order of the absorption length of the investigated mate-

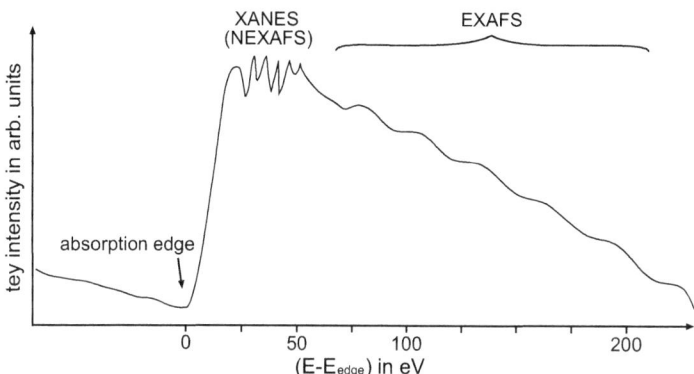

Figure 2.1: The picture shows the total electron yield, if one measures in transmission, the peaks in this graph correspond to dips in those spectra. Right above the edge the XANES region is located and about 50-100 eV after the edge the EXAFS wiggles start. When the energy of the incoming photons is tuned over the threshold of resonant absorption into empty states, a jump in the absorption cross section appears and an edge is visible in the XAS spectrum.

rial for the radiation energy used. If the target is a non-insulating solid and the absorption length of the radiation rather short, as for soft x-rays (e.g. a few 100 nm) one usually measures the emitted electrons (Electron spectroscopy) or the total electron yield (tey), which is the drain current of the target. Additionally the measurement of the emitted photons from the sample, the fluorescence yield (fy) is possible. For lighter elements (z<30) the dominant relaxation channel of core holes is non radiative Auger-decay (figure 2.2), leaving only a small portion for fy measurements. Principally it is possible with a thin solid target to record the x-ray absorption, tey and fy during the same measurement, while analyzing emitted electrons. Electrons that are emitted inside a material can be scattered and and partly absorbed by the material, which leads to a cascade of secondary electrons. This process is very similar for all materials (figure 2.3), therefore the sampling depth and the cascades of secondary electrons produced have very similar values for all metallic systems. However there have been evidences of a spin dependency of the mean free path [105] up to 30% for iron, which does not affect the measurements of this work since we did not measure the electron spin selectively.

The tey suffers a saturation effect, when the sampling depth of the electrons is in the order or greater than the shortest absorption length of the incident light in the energy region investigated. The effect occurs since the incident x-ray intensity at the target is a function of depth and of energy. The energy dependence is due to very distinct increases (absorption edges) in the absorption cross section of the incoming light at the energies related to the excitation from occupied into unoccupied electronic states of the target

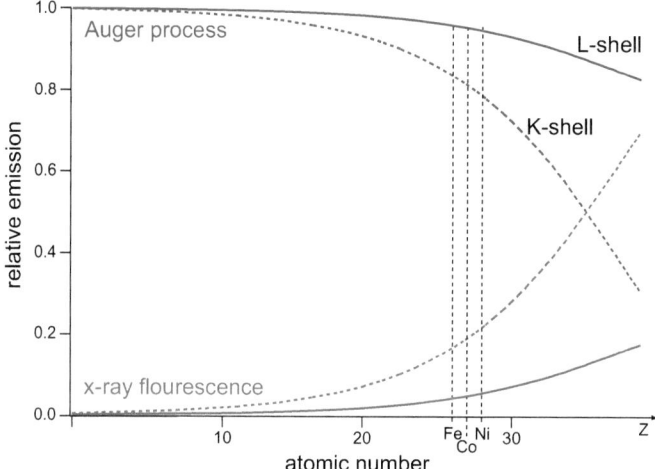

Figure 2.2: A scheme of the relative amounts of Auger electrons and x-ray florescence photons as relaxation channels for core hole excited states. With increasing mass of the nucleus the florescence yield takes over the Auger yield as the dominant core hole relaxation process. For 3d metals the Auger decay is by far the dominant process. The dotted lines represent K-shell hole relaxation processes, while the solid lines represent the relaxation of L-shell holes, as produced in the XAS measurements of this work. The figure is based on data from the X-Ray Data Booklet [139]

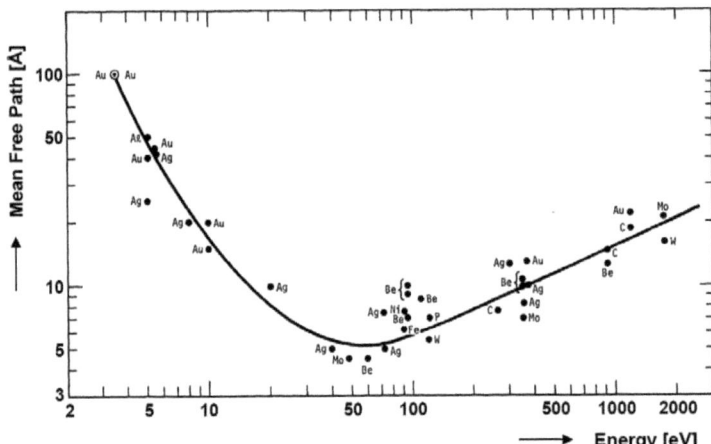

Figure 2.3: The electron mean free path follows a universal curve (figure from reference [128]). The mean free path depends on the energy of the electron and not on the material the electron is propagating in. In badly conducting materials or organic samples the mean free path is longer than predicted by the universal curve.

material. This decreases the absorption length of the light at the edge energy compared with lower energies. Measuring energy dependent across such absorption edges as done in this work one has to keep in mind that the absorption length of the light decreases just as the absorption cross section of the material increases. In the case of a strongly decreased absorption length at for instance the $3d$ L_3-edge, the assumption of a small sampling depth in comparison with the absorption length may not be valid any more. There is a detailed investigation of the electron yield saturation effect at the L-edges of Fe, Co and Ni by Nakajima et al. [99] and one work of K. Fauth specializing onto Co nanoparticles (1-30 nm diameter) [11] which includes the effect on spin and orbital magnetic moments. For the measurements performed in this work the saturation effect does not play a major roll, since all mass selected targets and iron films were sufficiently thin (one to three atomic layers) and all nickel films were similar thick and just used for relative calibration. Just the sum rule estimation of the orbital magnetic moment of the wet chemically synthesized nanoparticles was affected up to 15% (which was taken into account).

Apart from possible saturation effects measuring the **tey** is the measurement of all cross sections at the energy measured. This includes the absorption of substrates in case of thin layered samples. The measured signal is proportional to the incident x-ray intensity. J.Stöhr has discussed that in detail [132] and has been further inspected for cluster measurements by M.Reif [111]. Typically the measured spectrum will be divided by a spectrum simultaneously measured with a gold mesh (respectively the current on the last mirror reflecting the incoming radiation), important is that the ma-

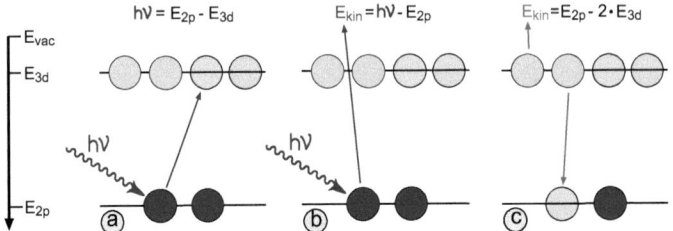

Figure 2.4: The picture shows the two cases of (a) resonant absorption and (b) non resonant absorption, as well as (c) the auger decay of the core hole which is left as a final state of both depicted absorption processes.

terial used for these spectra has no own absorption features in the energy region measured. This division removes possible glitches in the measured spectrum due to temporal instabilities of a storage ring causing fluctuations in the incident x-ray intensity.[1] As a second step the measured tey spectra are usually treated to remove the background caused by the substrate. A measurement of a clean background is needed for this. If the background is very smooth the spectrum correction can be done by subtraction, if the background has some explicit absorption features, the spectrum should be divided by the background spectrum measured. In case of the measurements in this work, the wet chemically prepared samples had a smooth linear background, that could be easily subtracted. The mass selected clusters had a background with explicit absorption features and therefore the measured spectra were divided by the background measurements. The typical procedure for this kind of measurement will be shown in the oncoming chapters 3.16 and 3.3.2.

---

1. This first normalization does not help, if the glitches are due to beam position shifts caused by storage ring instabilities, that lead to measurements on differently covered areas on the surface of the sample. In that case the measurement can not be used.

2.1.1 X-Ray Magnetic Circular Dichroism

X-ray magnetic circular dichroism (XMCD) is a specialized kind of XAS, for which one needs circularly polarized light to excite the core electrons of the target material. In a simple atomic one electron picture, the excitation probability[2] of left and right circular polarized electrons is not equal if a spin orbit coupling is present in the atoms, due to the so called Fano-effect. First calculated by U. Fano 1969 [39, 40] and displayed for the transition from $2p$ to $3d$ electrons at the L_2-edge (figure 2.6b) and L_3-edge (figure 2.6a). When absorbing a circularly polarized photon some of the orbital moment of the photon can be transferred via the spin orbit coupling to the electron spin. This effect leads to an effective spin polarization of the exited electrons. It is 75.0% at the L_2-edge and 62.5% at the L_3-edge. Due to the different spin-orbit coupling (l+s) at the $P_{3/2}$-state and (l-s) at the $P_{1/2}$-state the spin polarization at the two edges is differently oriented. If the target material is non magnetic, this leads to no visible effect in an absorption spectrum. If the material is magnetic the exchange splitting leads to a splitting of the valence band, lowering the energy for electrons with one spin orientation (majority electrons) and rising it for the other (minority electrons). The transition probability of electrons in a dipole transition are proportional to the unoccupied density of states (formula 2.2). If this density of states is different for different spin orientations and the electrons are spin polarized, the absorption spectra will change when the spin polarization is inverted due to the change of the helicity of the incoming light (figure 2.5).

---

2. actually the transition matrix element for a $2p-3d$ transition depends on the polarization of the exiting light

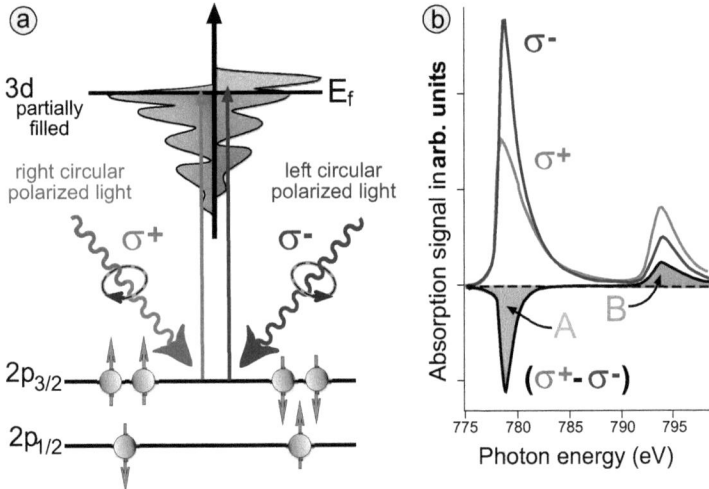

Figure 2.5: (a): A simple one electron two step picture of the resonant excitation process in a magnetic material, giving us an overview of the XMCD effect. In step one the circularly polarized photon excites a spin polarized electron (Fano effect) from the a $2p$ level; this spin polarized electron is used in step two to probe the spin polarized occupation levels of the $3d$ states. (b): Scanning over the $2p$ absorption edges of (here Co) with circularly polarized light the resonant absorption leads to non statistic absorption lines due to differently spin occupied $3d$ states. This effect is reversed when either the magnetization or the polarization is reversed. In this picture the polarization of the light is reversed between the blue and the red absorption curves. The difference of right and left circularly polarized scan is the dichroic signal and pictured as the black curve. The areas A and B of the difference curve can be evaluated by the XMCD sum rules to get access to the orbital to spin moment ratio.

To see this effect it is required that the magnetization of the sample has at least a component parallel to the incoming light. The inversion of the magnetization has the same effect on the absorption process as the inversion of the photon polarization. The dichroic behavior of magnetic materials has first been investigated experimentally with XMCD for the iron K-edge by G.Schuetz et al in 1987 [118].

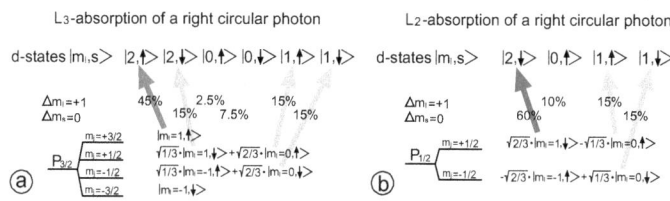

Figure 2.6: Absorption of circularly polarized photons lead to a different partial transmission probabilities as displayed above for the L_3 (a) and L_2 (b) edge. By excitation with circularly polarized light the selection rule demand that the change in orbital moment $\Delta m_l = +1(-1)$ for right (left) circularly polarized light. Spin flipping is generally forbidden, when exciting electrons with light, since the electromagnetic field does not couple with the spin, hence $\Delta m_s = 0$. The transition probabilities from all initial $2p$ states $P_{3/2}$ and $P_{1/2}$ into all the possible $3d$ final states in case of excitation with right circulalyr polarized light are shown. One can see that there are totals of 75% spin down excitations for the L_2 edge and 62.5% spin up excitations at the L_3 edge.

XMCD became more powerful as a tool with the derivation of the sum rules by Thole et al. [3, 4, 135, 136], which allow to directly estimate the orbital moment $m_l = -\frac{\mu_B}{\hbar}\langle L_z \rangle$ and the spin moment $m_s = -g_s \frac{\mu_B}{\hbar}\langle S_z \rangle$ of the material investigated using the formulas 2.3 and 2.4, using integrated absorption signals over the L_2- and L_3-edge region for left (σ^+), right (σ^-) circular and linear (σ^0) polarized light, the number of d-holes (n_h) and the z-component of the magnetic dipole operator 2.6 (T_z).

XMCD orbital sum rule:

$$\langle L_z \rangle = 2 \cdot n_h \cdot \hbar \cdot \frac{\int_{L_3+L_2}(\sigma^+ - \sigma^-)dE}{\int_{L_3+L_2}(\sigma^+ + \sigma^0 + \sigma^-)dE} \tag{2.3}$$

XMCD spin sum rule:

$$\langle S_z \rangle + \frac{7}{2} \cdot \langle T_z \rangle = \frac{3}{2} \cdot n_h \cdot \hbar \cdot \frac{\int_{L_3}(\sigma^+ - \sigma^-)dE - 2 \cdot \int_{L_2}(\sigma^+ - \sigma^-)dE}{\int_{L_3+L_2}(\sigma^+ + \sigma^0 + \sigma^-)dE} \tag{2.4}$$

total effective magnetization measured:

$$m_{tot} = m_{s(eff)} + m_l = 2 \cdot \frac{\mu_B}{\hbar} \cdot \left(\langle S_z \rangle + \frac{7}{2} \langle T_z \rangle \right) + \frac{\mu_B}{\hbar} \cdot \langle L_z \rangle \tag{2.5}$$

For a precise estimation one must separate the L_3 and the L_2 contribution of the spectrum, which is fairly easy for the heavier $3d$ elements, but can proof a hard task for light $3d$ elements since the spin-orbit split increases with the nuclear charge of the atom. Additionally in case of up excitations one needs to know the exact amount of d-holes in the material, for which no straight forward determination procedure exist, therefore the results in this work are presented mainly as magnetic moments per d-hole. Finally one needs to have the knowledge of the asphericity of the charge density of the valence states $\vec{T}$ (formula 2.6).

$$\vec{T} = \vec{S} - 3\hat{r}(\hat{r} \cdot \vec{S}) \tag{2.6}$$

The asphericity is often assumed to be zero, which is usually a good assumption for cubic undistorted systems. Quite often the distortions are responsible for the magnetization of the system, as the effect of perpendicular magnetic anisotropy (PMA), which is present in the magnetic thin film systems used as magnetic sublayers for the mass selected clusters in this work. For most 3d-metallic systems especially for iron, cobalt and nickel it has been shown that the total error is less than 10% if $\vec{T}$ is included in the total error [25, 102, 152, 153]. Theoretical calculations of Ederer et al [36] with Linear muffin-tin orbital method (LMTO) and full-potential linearized-augmented-plane-wave method (FLAPW) have shown that $\vec{T}$ in $Co_m Pt_n$ multi layer system will increase strongly at the interface. They claim T_z not to be negligible for the interface layers in flat layered systems.

For practical application of the sum rules a measured tey spectrum has to be treated as described above to remove the features inflicted due to instabilities and background. If that is done the resulting spectrum still contains all contributions due to all partial absorption cross sections of the sample material in the measured energy region. The sum rules explicitly cover only the 2p to 3d excitations, therefore all other contributions must be removed from the measured spectra, before applying the sum rules. These contributions are removed by subtraction of a hyperbolic step function with one step at the L_3-edge and one at the L_2-edge with a step height ratio of 4:2, following the multiplicity 2j+1 of the edges. There is some discussion about the origin of this step. If the step would originate only from p- to s-state transitions, it should be much smaller. Only about 5% of the total absorption cross section is related to 2p to 4s transition [33]. A small amount of s-d hybridization could boost the absorption cross section enough to enlarge this step to the measured amount. Measurements of S.Liu et al. on the Valence states of free size selected Co and Ni clusters show an evolution of separated s and d states towards a mixed *s-d*-state for a cluster size as small as 20 atoms [86], but so far no overall accepted explanation for this boosted cross section is given in literature. Nonetheless the described practice has proven to work well and is used by all experts in this field. Little discussion is about the position of the step that is defined by the fermi level of the system. In experiments of this kind it is difficult to measure the fermi level of the system, hence it is common practice to vary the position of the step function slightly until a result is reached in which the peak properties of the absorption peaks are changed as little as possible by subtracting the step function. Typically the steps are anchored at the maximum position of the absorption peaks or as in this work the inflection point of the derivative of the absorption peaks (which should principally represent the fermi level of the system). Finally the resulting spectrum consists only of the contributions from $2p_{1/2}$ and $2p_{3/2}$ to empty *d*-state transitions.

The dichroic signal can be evaluated without subtraction of the step function and usually even without subtraction of any background or slope. The dichroic signal only gives access to the ratio of orbital and spin magnetic moment (formula 2.7). One has to keep in mind that this is the ratio of orbital and effective spin moment, which includes the contribution of $\langle T_z \rangle$.

$$\frac{m_l}{m_{s(eff)}} = \frac{2}{3} \frac{\int_{L_3+L_2}(\sigma^+ - \sigma^-)dE}{\int_{L_3}(\sigma^+ - \sigma^-)dE - 2 \cdot \int_{L_2}(\sigma^+ - \sigma^-)dE} = \frac{2}{3}\frac{A+B}{A-2\cdot B} \quad (2.7)$$

For this ratio errors due to uncertainties mentioned above drop out and allow the ratio to be determined more precise than either of the involved magnetic moments. For absolute values of either spin or orbital magnetic moment it is nevertheless unavoidable to undergo the complete background treatment. While the spin moment can be estimated quite good in most cases of negligible $\langle T_z \rangle$ contributions, the orbital moment is generally underestimated by the sum rules. This error is in the order of 25% of the final value for Fe and Co [51, 127].

Often the dichroism normalized to the L$_2$-edge is displayed to visualize changes in systems, these graphs display a variation of the orbital to spin magnetic moment. Obvious it can not represent one of the magnetic moments, since their values depend on the normalization by the whiteline spectrum, which would be canceled, when normalizing to the L$_2$-edge (area B in figure 2.5b) of the dichroism spectrum. Normalizing the area B to one, the area A will be scaled by the same factor and represents after this normalization procedure, the former ratio of A/B. This ratio can be related to spin and orbital moments by solving equation 2.7 for this ratio:

$$\frac{m_l}{m_{s(eff)}} = \frac{2}{3}\frac{A+B}{A-2\cdot B} \qquad (2.8)$$

$$\Rightarrow \frac{A}{B} = -\frac{2\cdot m_{s(eff)} + 6\cdot m_l}{2\cdot m_{s(eff)} - 3\cdot m_l} \qquad (2.9)$$

$$\Rightarrow \frac{A}{B} = -1 - \frac{9}{2}\frac{m_l}{m_{s(eff)} - \frac{3}{2}m_l} \qquad (2.10)$$

Typically $m_{s(eff)}$ is 5-10 times larger than m_l, hence in good approximation of $m_{s(eff)} \gg \frac{3}{2}m_l$ the ratio of the areas A and B represent the ratio of $m_l/m_{s(eff)}$:

$$\frac{A}{B} \approx -\frac{9}{2}\frac{m_l}{m_{s(eff)}} - 1 \qquad (2.11)$$

The displayed ratio is not exact and has an offset, but trends of m_l/m_s can be nicely visualized by this form of representation.

For the experiments in this work the magnetic fields were chosen to be aligned parallel to the incoming photons, to maximize the measurable effects of XMCD. To achieve this in addition to a maximum light flux per sample area, during the measurements of the mass selected particles, with coverage of less than 5% of a ML, surfaces with magnetization perpendicular to the surface normal were used.

2.1.2 Thermal Desorption Spectroscopy

Dealing with thin film systems or using soft landing processes the calibration of evaporators and the investigation of the cushion layers for the landing process are important. In case of this work all of that has been done using the titration method thermo desorption spectroscopy (TDS). TDS is an extremely surface sensitive technique, that probes only the interaction between the topmost surface layer and the adsorbed gas. Adsorbates can bind chemically or stay chemically adsorb or physically adsorb on a surface. Chemical bonds should be avoided if non destructive investigations are preferred. If the samples are cooled to low temperatures noble gases as argon, krypton or xenon are best suited for a chemically non interacting titration experiment on a surface. For TDS one produces a clean surface to investigate, freezes out 2-3 layers (if possible) of the titration gas (for this work xenon) and then slowly heats up the sample while measuring the partial pressure of the desorbing titration gas. A typical desorption process is shown on the left side of figure 2.7.

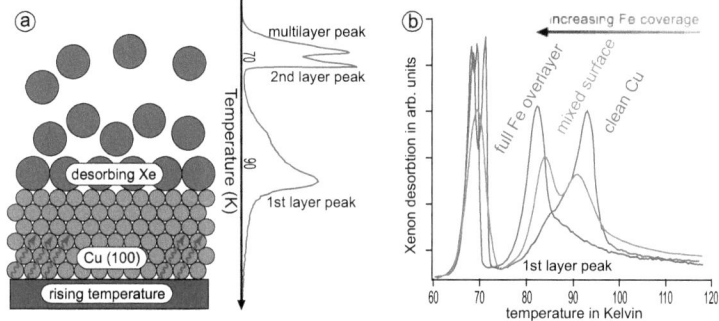

Figure 2.7: An overview over the titration method of thermal desorption spectroscopy (TDS), using the titration gas xenon and a copper surface: (a) After freezing several layers of xenon onto the clean surface the sample temperature is slowly increased, while the partial pressure of the xenon is measured. The combination of temperature and pressure measurement give a curve that show as a multi layer peak, a second and a monolayer peak. (b) TDS for the case when the substrate is changed between succeeding measurements. The iron coverage increases from nothing to a complete layer. The position of the monolayer-peak shifts with the amount of iron, until a full layer of iron is on top of the substrate. The multi layer peak is unaffected by the change of the surface.

The Van-der-Waal's interaction between the noble gas and the surface is stronger than between two noble gas atoms and the second layer of the noble gas is still noticeably stronger attracted than the following layers, producing a spectrum containing from the low temperature side a multi-layer-peak (3rd layer and above), a second layer peak and a monolayer peak. If the substrate is changed the temperature position of the multi layer peak does not change, but the interaction between the lowest layer of the noble gas and the surface will be different, which leads to a shift of the temperature position of the monolayer peak. For evaporator calibration one takes a data set with TDS measurements of increasing coverage from 0 to 1 layer. One can see the movement of the first layer peak (right side of figure 2.7). For evaluation of frozen cushion layers as in the case of the soft landing process one can integrate the areas of the peaks and by knowing that the monolayer peak contains the amount of gas necessary to cover the sample, estimate the total coverage.

2.1.3 Soft Landing

The process of depositing clusters from the gas phase on surfaces is always connected with the risk of fragmenting the cluster or even implanting them into the substrate.

Molecular dynamic simulation and experiments have shown, that clusters of only 13 atoms may still fragment when the kinetic energy of the cluster is even less than 1 eV/atom [26, 27, 64, 144]. When keeping the energy per atom constant, the risk of fragmentation increases with the size of the cluster and the kinetic energy. Hence in the experiments the total kinetic energy per cluster was kept constant, reducing the kinetic energy per atom with increasing cluster size. The use of several layers of noble gases can strongly reduce the fragmentation of the clusters upon impact, by dissipating the kinetic energy [20, 21, 43, 79]. The largest cluster deposited on the surface in this work was only 4 atoms in size and the highest kinetic energy of the entire clusters was below 1 eV/cluster. In these cases the risk of fragmenting is already low, but to further reduce the chances, the well established procedure of soft landing was used. Measurements of S.Fedrigo (figure 2.8) of mass selected Ag^{2+} have shown that in the case of low energetic clusters the effect of argon

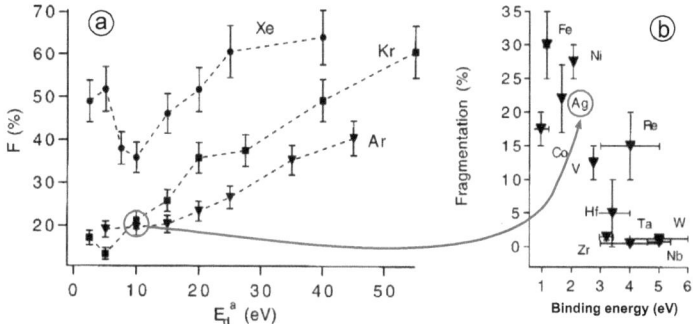

Figure 2.8: The fragmentation probability of Ag$_2$ clusters in respect to the kinetic energy for different rare gas matrices (a). The fragmentation probability of Dimers of different materials by deposition in an argon soft landing matrix with 10 eV kinetic energy as a function of the intra dimer binding energy is shown in (b) (figures from reference [13]).

and krypton layers are equally effective for the soft landing. Additionally they discovered that a higher binding energy makes clusters usually more stable, only Cobalt was in spite of its rather small inter dimer binding energy of 1 eV relatively stable.

2.1.4 STM

One very interesting aspect would be to investigate deposited clusters with an scanning tunneling microscope (STM). There has been much planning done to realize STM investigation of in situ prepared mass selected deposited clusters in our group. It would be very interesting to investigate the surface and the deposited clusters with a visual method. The interpretation of the measured data could be much improved by the knowledge of the geometric orientation of the clusters on the surface, as well as the actual appearance of the surface itself before and after cluster deposition.

There is very nice work done by the group of group of R.Berndt (figure 2.9). Mn_1 to Mn_4 was produced by STM manipulation of Mn adatoms on a smooth Ag(111) surface.

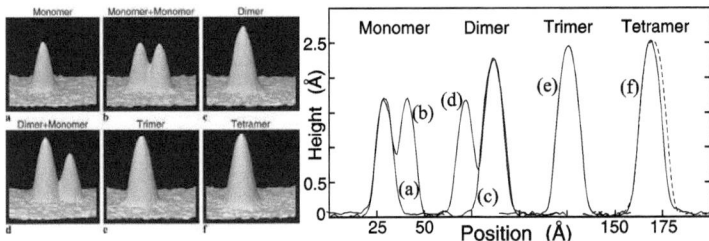

Figure 2.9: STM measurements of metal clusters upon metal substrates are very difficult. The Mn-clusters were in situ prepared by STM manipulation technics on a very smooth Ag(111) surface by R.Berndt et al. (figure from reference [74]). It already proves hard to distinguish Dimer, Trimer and Tetramer and thus demonstrates the challenge one will have to rise to, in order to decisively identify metal clusters on rough surfaces after deposition.

The changes from one size to another are unfortunately very small, hence it will be quite a way to go, until larger clusters on a rougher surface will be equally nicely measurable. The stability requirements of a STM to achieve this resolution is too high to be able to perform this with a traveling experiment at a synchrotron. The usual manipulators used for synchrotron experiments allow by far too much movement of the sample at the possible location of the STM, therefore either a special transfer system must be developed or more likely a special STM would have to be constructed, that allows the sample preparation and cluster deposition inside the STM. These are very challenging experimental difficulties to be solved in order to lift the secret of the geometry of the deposited clusters by means of STM.

2.2 Material properties

2.2.1 3d transition metals

The 3d transition metals are investigated in experimental an theoretical physics intensively. They are relatively common and stable, are often used in technological applications and that almost during the entire history of human civilization. They show interesting properties such as ferromagnetism in the case of Fe, Co and Ni even as bulk material and at room temperature. In this section some spectroscopically relevant properties of 3d transitions are introduced, first for 3-dimensional bulk material, then reducing the dimensions to 2 and below, as it is the case for thin films or small islands (as deposited clusters).

While the outer 4s shell is filled for most 3d transition metals in the ground state, the 3d shell is subsequently filled from Sc_{21} to Zn_{30} giving rise to the very unique properties of these elements. The

energy	interaction
$\sim 5eV$	3d-hybridization (3d bandwidth)
$\leq 2eV$	multiplet splitting (electrostatic)
$\sim 1eV$	magnetic exchange interaction
$\sim 0.1eV$	crystal field splitting
$\sim 0.05eV$	spin orbit coupling

Table 2.2: This table lists the typically occurring interactions in bulk metal and the energies involved for 3d metals.

3d transition elements are quantum mechanically in the region of so called intermediate coupling between LS and jj coupling, which is important for theoretical model calculations. The outer weakly bound electrons can still be treated as it is done for the light elements of the periodic table, where S and L are good quantum numbers, the more strongly bound deeper lying electrons on the other hand already experience relativistic effect. Experimentally the electronic structure can be investigated either by probing the filled $3d/4s$ density of states (DOS) or the empty $3d/4s$ states. The empty states can be probed by optical excitation using the above described XAS. To first order in a one electron picture, which we use for the explanation of the magnetic dichroism, ignoring any spin-orbit coupling of the initial state $2p^63d^n$ and any electron core hole interaction in the final state $2p^53d^{n+1}$, one would expect an statistical branching ratio (BR) of 2 between the integrated absorption from $2p_{3/2} \to 3d$ and that of the $2p_{1/2} \to 3d$ excitation. Experiments of Fink et al. in 1985 [47] have shown that the BR is about 0.7 for Ti and rises to 2.3 for Co (figure 2.10a). In 1988 Thole and Van der Laan [137] discussed the effects that using LS or jj coupling has on the Hatree Fock (HF) calculated spectra, coming to the conclusion that in the case of nickel for the $M_{2,3}$ absorption edges almost undisturbed LS coupling can be used, while the $L_{2,3}$ edges show some 70% jj coupling behavior (figure 2.10c). The BR varies with the type of the coupling.

In 1998 Schwitalla and Ebert [119] addressed the problem of the BR of $3d$ transition metals with a fully relativistic time dependent density functional calculation at the $L_{2,3}$ edges (figure 2.10b). Their results showed that the electron core hole interaction in the $2p^53d^{n+1}$

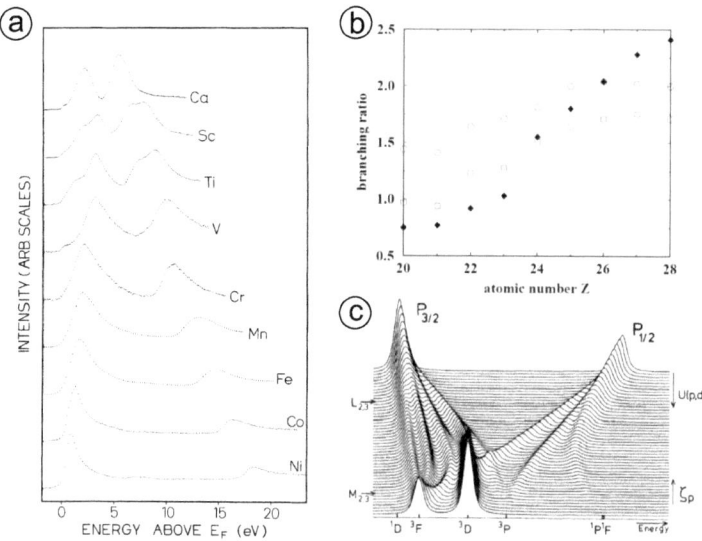

Figure 2.10: The statistical branching ratio of the L_3/L_2-edge is 2:1 and as shown in (a) by Fink et al. (figure from reference [17]) for bulk material smaller for early and higher for late $3d$ metals. These measurements could be theoretically confirmed with fully relativistic TD-DFT calculations by Schwitalla et al. (b) (figure from reference [119]). Thole and Van der Laan could show with HF calculations (c) (figure from reference [143]) that the L-edge absorption of $3d$ transition metals behave 70% as jj-coupling induced and only 30% as LS coupling induced, which states that relativistic effects must be included when dealing with $3d$ transition metals.

final state causes an intermixing between the L_2 and L_3 partial spectra, thus affecting the branching ratio. The stronger the core hole interaction the more the BR is shifted in favor of the L_2-edge. Their calculations produce the right trend, but still overestimate the BR of early and underestimate that of the late $3d$ transition metals. Scherz et al. [116] extended this and included exchange correlation effects. By allowing the core hole to affect the spin orbit coupling, the experimental data for bulk samples could be reproduced by calculations for Ti, V, Cr and Fe.

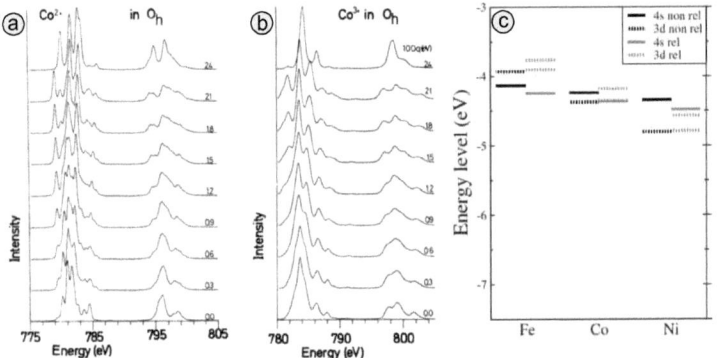

Figure 2.11: (a) and (b) (figures from reference [28]) show atomic multiplet calculations of F.M.F. de Groot for Co^{2+} and Co^{3+} as a function of the lattice parameter (10 Dq). The spectral differences between Co^{2+} and Co^{3+} are strong enough to identify the different contributions in absorption spectra measured and to estimate roughly the effective lattice parameter present at a measurement. F.Aguilera-Granja et al. calculated 2008 the energy levels of $3d$ and $4s$ states for Fe, Co and Ni (c)(figure from reference [1]) with and without relativistic effects, which lead in the case of cobalt to a more filled $4s$ state s-d-band when including relativistic effects.

Metallic $3d$ systems produce rather simple $2p \rightarrow 3d$ absorption spectra, as can be seen in figure 2.10a. If the 3d metals oxidize or are in other covalent bound states, the band structure of the 3d-band changes to an orbital molecular structure and the binding energy of the 2p level shifts due to changed screening [90]. The $2p \rightarrow 3d$ absorption spectra consequently move in the energy position and show some fine structure, as it is calculated for the two different lattice sites of Co in Co_3O_4 in dependence of the crystal field parameter (10Dq) by F.M.F.de Groot et al. with atomic multiplet theory (figure 2.11 a+b). The spectra of Co^{2+} in CoO looks as the calculated spectra for the 2^+ position, when assuming the same symmetry. The position of the L_3-edge of the Co^{2+} has it's maximum at the energy position of metallic cobalt, while the absorption edge is shifted roughly 4 eV for the Co^{3+} spectra. In addition the line shape of the two spectra are distinctively different. These changes can be used to analyze the composition of a sample. In experiments this effect is often exploited to exclude the presence of a specific oxide, as Co_3O_4, as in the experiments of this work.

Most calculations and investigations are performed for bulk materials and atoms first due to the high symmetry of those systems and later extended to thin films and clusters. Depending on the question addressed the size at which the material changes it's atom like properties to bulk values is often different. The magnetic properties of free Fe,Co and Ni clusters reach bulk values at 400-700 atoms/cluster [17], some electronic properties seem to change earlier, as the 4s-3d hybridization that is present in bulk material and not for atoms. Figure 2.12 shows some photoelectron spectra of S.-R.Liu et al [85, 86] of free size selected Co_n clusters. The mixed s-d-band appears already at n$\geq$20 (for nickel it is already at half the size).

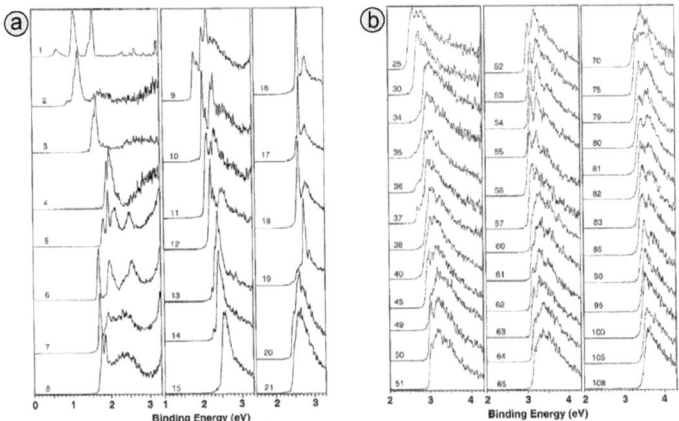

Figure 2.12: Photoelectron spectra of free size selected Co-clusters of S.-R.Liu et al.(figures from reference [85]) show nicely the transition from atomic s and d orbital's towards a s-d-band like structure at cluster size 20.

The $3d$-band of clusters with sizes of 1 nm diameter ($\approx$ 100 atoms) is already bulk like, and thus the spectral effects due to oxidation mentioned above, can be used for Co nanoparticles. How deposited clusters on metals evolve in their electronic states has so far not been investigated, hence more caution is advised.

The magnetism of the $3d$ transition metals is dominated (over 95%) by the $3d$ valence electrons. Bulk Fe, Co and Ni are usually pictured in the Stoner Model, splitting the valence band in one spin up band (minority spin density $N_\uparrow$) and one spin down band (majority spin density $N_\downarrow$). The majority band is energetically lower and hence more occupied, which leads to a resulting permanent spin moment of the material.

2.2.2 Cobalt-Platinum alloy

The high complexity of the electronic structure of CoPt alloyed system has been and still is a very challenging task for all theoretical calculations, therefore there are some calculations for bulk materials, very few for free clusters and none (to the best knowledge of the author) for clusters in contact to surfaces. Nontheless CoPt alloys have been a materials of high interest over the last 20 years, especially because of the outstanding magnetic properties of those alloys (high magnetic anisotropy and high coercivity) and the good corrosion resistance [11, 84, 112].

Vapor deposited $CoPt_3$ and $CoPt_2$ films have shown that $CoPt_3$ crystalizes in $L1_2$-phase and that $CoPt_2$ is more complicated at low temperatures being at the two phase region with both phases ($L1_0$ and $L1_2$) present. Shapiro et. al. believe to have seen Co clustering during the deposition if the deposition temperatures were below 450 °C [121].

MBE deposited $CoPt_3$ 500Å thin film XMCD measurements show out of plane magnetic anisotropy for orbital moment [53].

CoPt films with different stoichiometric ratios were investigated by Weller et al. [117] with Kerr spectroscopy. 25 nm films were created using electron beam evaporation at temperatures of 100 °C to 250 °C and as expected a strong PMA was observed and the remanent magnetization increased with the evaporation temperature.

Theoretical calculations using a fully relativistic muffin-tin orbital method of CoPt and $CoPt_3$ have been performed by Galanakis et al. [19] (figure 2.13a), assuming CoPt to be in the $L1_0$-phase and $CoPt_3$ in the $L1_2$-phase (figure 2.14). Assuming a number of d-

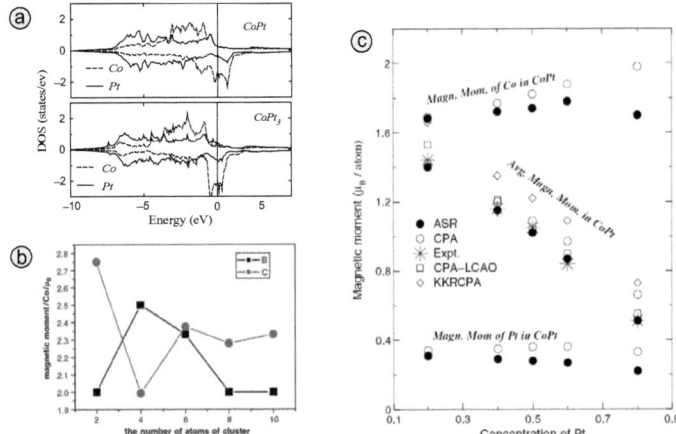

Figure 2.13: (a) FR-MTO calculations as of Galanakis et al. (figure from reference [49]) for CoPt and CoPt$_3$ show that Pt has very little empty states at and above the fermie edge, while Co has many and especially for CoPt$_3$ giving rise to an enhanced magnetic moment for Co. (b) GGC DFT calculations of pure Co$_n$ clusters (black squares) and (CoPt)$_{n/2}$ cluster (gray filled circles) (figure from reference [44]) show a principally enhanced magnetic moment for Co in the alloy clusters. (c) An ASR TB-LMTO method calculation for bulk CoPt-alloys of D.Paudyal et al. (figure from reference [106]) confirm the trend of rising magnetic moments at the Co sites with increasing Pt contend of the alloy, as to be expected by the DOS shown in (a). The total magnetic moment of the alloy decreases due to the much smaller moments of the Pt, which is consistent with the experimental data cited in the article.

holes for CoPt (CoPt$_3$) of n_d = 2.651 (2.628) their results were fairly similar for the spin moment m_s = 1.8 (1.9) μ_b, but differed almost a factor of two for the orbital moment m_l = 0.095 (0.055) μ_b. The ratio of orbital to spin magnetic moment (which is the most accessible for experimentalists) was m_l/m_s = 0.053 (0.029). A very similar trend was calculated by Sipr et al. in 2008 [115] in a fully relativistic calculation, using the dynamic mean field theory to include many-body effects. The ratio of orbital to spin magnetic moment dropped from 0.139 for CoPt to 0.094 for CoPt$_3$. Even the values of Sipr et. al are still below any experimental references to be found in literature. Sipr et al. reported that the increase of Pt concentration is suppressing the Co orbital moment, since Platinum has very little empty states above e_f (the calculated DOS was very similar to the displayed results of Galanakis et al. in figure 2.13a) and thus leaves the Co minority states less hybridized. Generally the mixing of Co and Pt atoms leads to a reduction of the quenching of orbital moments, compared with bulk cobalt. Decreasing the long range order of CoPt$_3$ Sipr et al. found almost no effect on the orbital moment, but for the Co atoms the average amount of nearest neighbor Co atom increases from zero to a maximum of three and an increase of the spin moment of up to 5%.

In 2006 Feng et al. published their generalized gradient corrected spin density functional results of (CoPt)$_n$ clusters with ($1 \leq n \leq 5$) calculations. They calculated the ground state properties of different geometries, the DOS and the magnetic properties of those clusters and the magnetic moments of the involved atoms. Their results have shown an continuous increase of the total magnetic moment per (Co-Pt)-pair with increasing cluster size, but not for the

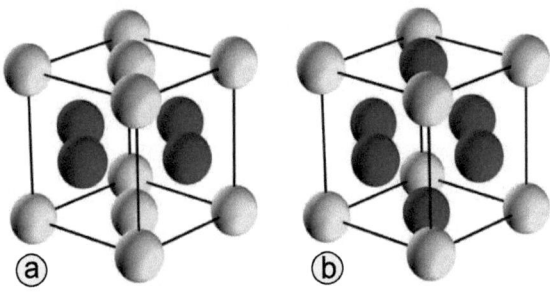

Figure 2.14: Both structures are fcc structures. In the $L1_0$-structure (a) there are alternating planes of atoms of different type in [100] direction and thus an equal amount of atoms of the two different types. In the $L1_2$-structure the ratio of the two different atom types is 1:3, hence every second plane in [100] direction is a mixed plane of atoms of both types.

magnetic moment of the average Co atom in those clusters (figure 2.13b).

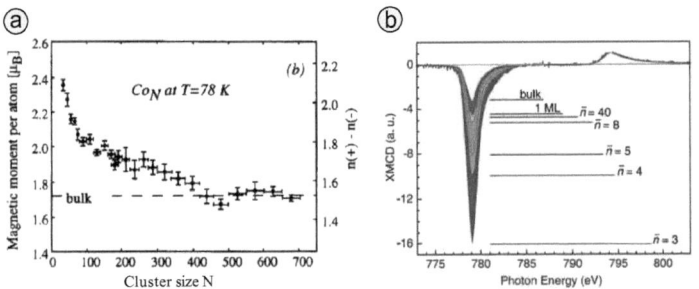

Figure 2.15: (a) Measurements of free mass separated Co-cluster (figure from reference [17]) Have shown decreasing magnetic moments per cobalt atom with cluster size, reaching bulk value at 400 atoms per cluster. (b) Deposited mass separated Co clusters on Pt(111) (figure from reference [50]) displayed a very high orbital to spin moment ratio (up to 400% of the bulk value) and thus raising hope for extremely high magnetic anisotropy

Previous experiment on free Co cluster [17] have shown that there are strong size dependencies up to 50 atoms/cluster and that still up to 400 atoms/cluster the magnetic moment is well above the bulk value (figure 2.15a). A 400 atom cluster has 2-3 nm diameter and a surface to volume ratio of roughly 1/2. Co clusters in the range of 3 to 40 atoms/cluster were deposited on a Pt(111) surface and displayed an enormous ratio of orbital to spin magnetic moment, up to 4 times that of bulk cobalt (figure 2.15b).

2.2.3 Magnetic thin films

Thin film often show very different magnetic properties compared with the bulk material. The most obvious reason for that is breaking of the symmetry at the surfaces of the film, which implies a reduced coordination of the atoms. Less d-d hybridization leads to a narrowing of the d-band resulting in an increase of the spin magnetic moment. At the surface of a bulk material or in a thin film the quenching of the orbital moment that occurs in highly symmetric systems is strongly weakened.

Furthermore the density of states at the fermi-level can be higher due to surface states and the consequential further narrowing of the d-band, which also increases the orbital moment. In addition in very thin films the pseudomorphic growth on the substrate material will cause lattice strain, which again breaks the symmetry and gives rise to enhanced orbital moments (no or less quenching than in the bulk).

The orbital moment couples to the lattice field due to the anisotropy of the crystal field. The spin couples only through spin orbit interaction with the orbital moment to the lattice and is therefore almost

free (isotropically oriented).
Surface shape anisotropy favors an in plane magnetization and is usually far to small to explain the total magnetic anisotropy energy (MAE). Small distortions of the lattice of 5% may increase the MAE energy by $10^2 - 10^3$ [10]. The lattice distortion is usually stronger perpendicular to the surface giving rise to an increased perpendicular magnetic anisotropy.

The magnetic anisotropy of a film can be expressed in a simple picture as the sum of the volume (K_V) and the surface (K_S) anisotropy and the film thickness (d) in formula 2.12:

$$K = K_V + \frac{K_S}{d} \quad (2.12)$$

The influence of the surface decreases with the film thickness. Surface anisotropy tends to favor an out of plane magnetization, while the volume anisotropy favors in plane, except in some cases where a lattice distortion is present. Usually thin films have therefore a small thickness region of out of plane magnetization and with increasing thickness a spin reorientation transition (SRT) occurs in which the magnetic easy axis of the film tilts into the surface plane. More detailed information about magnetic anisotropy effects and their origin for especially transition metal thin films can be found in [11, 22, 114, 143].

Ni/Cu(100)

The system of Ni on Cu(100) is very docile for experimentalists. The very slight difference in lattice constants of nickel ($a_{Ni} = 3.52$ Å) and copper ($a_{Cu} = 3.61$ Å) imply a very slight lattice mismatch ($\eta_{Cu}^{Ni} = (a_{Cu}\text{-}a_{Ni})/a_{Ni}$), which again causes relatively little lattice strain and makes the changes in the induced magnetic properties very gradual. Ni/Cu(100) does not alloy at temperatures below 400 K, although first signs of alloying have been reported for temperatures of roughly 300 K [57]. Nickel grows pseudomorphic up to about 20 atomic layers on Cu(100) then gradually changing to nickel bulk structure. In the pseudomorphic range the nickel experiences a lateral expansion of approximately 2.5% and a contraction in the inter layer distances of about 3.0% [16, 60, 72, 100, 109] growing in Ni-fct, no longer quenching the orbital moment. The lattice distortion gives rise to a perpendicular magnetic anisotropy (PMA) between 7 and 40 ML [101] (figure 2.16). The perpendicular magnetization is almost constant over the thickness region between 10 and 30 layers and the lattice parameters change very little. The SRT takes place by small in plane domain formations and growth, hence for the experiments in this work only the magnetization of the nickel thin films was extracted from the data (using the XMCD sum rules) to calibrate the magnetization of the deposited clusters. The substrate has been used successfully used for spectroscopic measurements of mass selected clusters [80, 81, 154].

Fe/Cu(100)

Bulk iron crystalizes at room temperature in body-centered cubic (bcc) structure (α-Fe) with a lattice parameter ($a_{Fe}^{bcc} = 2.87 \text{\AA}$) (table 2.1). The high temperature (1185-1667 K) fcc-phase (γ-Fe) of bulk iron has a much larger lattice parameter ($a_{Fe}^{fcc} = 3.52 \text{\AA}$) [56].
In thin films on Cu iron grows pseudomorphic and does not intermix with the substrate up to 300K. This layer system Fe/Cu has been intensively investigated in the last 20 years. Iron thin films on copper

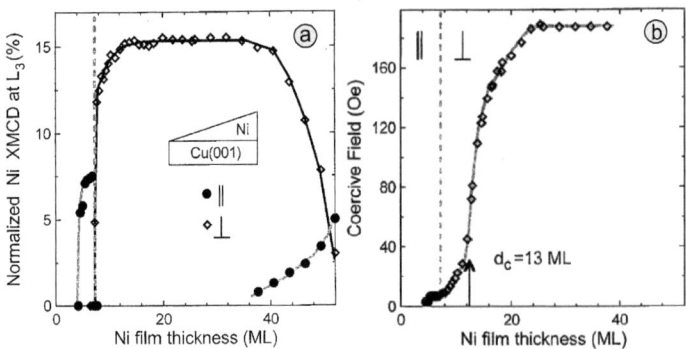

Figure 2.16: Nickel thin film on Cu(100) magnetization measurements (figures from reference [101]) shows an almost constant perpendicular remanent magnetization for thicknesses between 20 and 40 atomic layers.

have either been prepared with a rather smooth surface at room temperature or with increased surface roughness at temperatures below 100 K [15, 48, 96, 126, 130, 155, 156]. Most measurements lead to the picture that Fe thin films on Cu(100) grow in three thickness regions. The first 4 layers (region I) grow pseudomorphic with a slight vertical distortion as tetragonally distorted Cu-fcc (fct) structure, giving rise to a PMA between 2 and 4 fe-layers with a Fe-fcc high spin state. Layer 5 to 10 (region II) grow in nearly undistorted γ-Fe-fcc structure with slightly elongated inter layer spacing, however with a Fe-bcc like reconstructed over layer, while the magnetization tilts into the surface plane. The magnetization of the topmost layer remains out of plane. Above 10 atomic layers (region III) the Fe-film assumes the native bcc-Fe structure with the bcc-[110] plane parallel to the surface and in plane magnetic easy axis. Spin polarized metastable-atom deexcitation spectroscopy which is only sensitive to surface side of the topmost layer [76] (figure 2.18b) has shown that the magnetization of the topmost iron layer on Cu(100) grown at 300 K remains perpendicular to the surface at least up to 8 Fe-layers, although the overall magnetization measured with Kerr ellipticity tilts in plain above 4 Fe-layers. Earlier measurements with spin polarized secondary electron spectroscopy [101] had suggested a PMA for iron films on Cu(100) grown at 125 K between 1 and 6.1 Fe layers. XMCD measurements in an angle of 67° to the surface normal of iron thin films on Cu(100) have shown [117] strongly enhanced spin moment for iron (high spin state) in region I and a low spin state in region II. Our XMCD measurements perpendicular to the surface at 40 K confirmed that with the change of structure between region I and II the total net magnetization in perpendicular direction diminishes. Cobalt coverage of 10-20% of a ML as well as the increased surface roughness of thin films prepared at lower temperatures lead to an earlier SRT [122] (figure 2.17).

With increasing iron film thickness the topmost two iron layers remain perpendicular magnetized, while starting at the third layer, the films vary strongly, showing no magnetization at higher temperatures and developing a spin density wave structure at 130K (figure 2.19), as could be shown with depth resolved XMCD measurements by K.Amemiya et al. [5, 6]. They investigated a 7 ML film Fe/Cu(100) and established a picture of a spin density wave state with increasing anti-ferromagnetic inter layer coupling of the Fe-film at decreasing temperature, leading to a decrease of the XMCD signal with decreasing temperature.

Iron thin films as substrate for mass selected clusters was first used for depositing Chromium clusters [111] and has proven to work

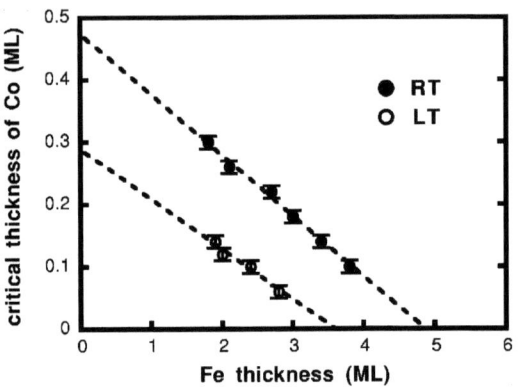

Figure 2.17: The spin reorientation transition of iron thin films occurs at much lower thicknesses, if the thin film is covered with cobalt, as experiments of Shen et al. (figure from reference [122]) have shown. The two abbreviations RT and LT stand for room temperature and low temperature grown films.

well, but being experimentally much more challenging than the Ni/Cu(100) system and in the data evaluation process especially at the background treatment. Iron thin films show temperature and thickness dependence, while the surface layers stay perpendicular magnetized for all Fe/Cu thin film systems. Clusters couple via exchange coupling to the uppermost surface layer and since that is for Fe/Cu(100) always magnetized perpendicular to the surface plane, the clusters show their maximal magnetic moment in a perpendicular direction.

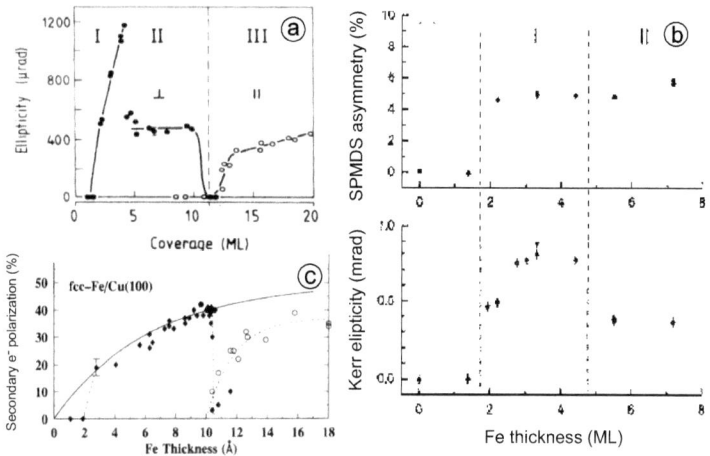

Figure 2.18: Iron thin film on Cu(100) shows a perpendicular magnetization in a thickness rage of about 2 layers (2-4 or 3-5, depending on the source). (a) The magneto-optical Kerr effect (MOKE) measurements (figure from reference [138]) show 3 different magnetic regions (I,II and III), where region I represents a high spin state of the iron with strong out of plane magnetization, region II a low spin state with still out of plane magnetization and region III a low spin state with in plane magnetization. (b) MOKE and spin polarized metastable deexcitation spectroscopy (SPMDS) measurements confirm the regional structure (so do XMCD measurements not shown here), while the top-layer sensitive SPMDS proves the upper lay er to have equal perpendicular magnetization in region I and II (figures from reference [76]) . (c) Spin polarized electron spectroscopic investigations (figure from reference [104]) too confirm the regional change, but show in plane and out of plane magnetization in region II.

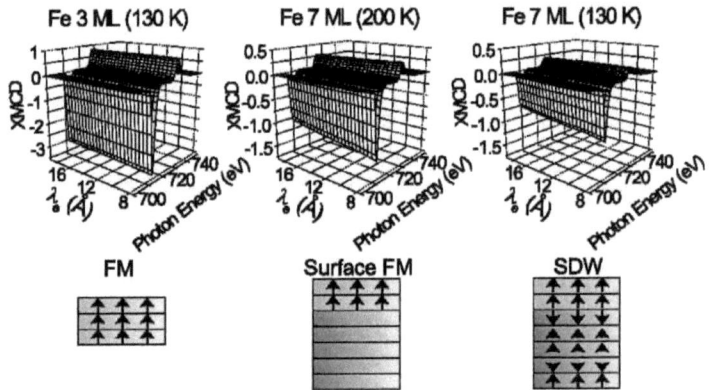

Figure 2.19: The magnetic structure of iron thin films on Cu(100) is rather complicated above region I (up to 4 layers). Depth sensitive XMCD measurements (figure from reference [5]) gave information that confirmed the idea of a Spin density wave state that establishes in region II (up to 10 layers) underneath the 2 upper iron layers, that keep a fixed perpendicular magnetization. This spin density wave (SDW) state disappears at medium temperatures (200 K) and reappears near room temperature. The model agrees fine with all measurements so far (e.g. figure 2.18)

Chapter 3

Experimental Setup

This chapter will introduce the different experiments performed in this work, the sample preparation methods and experimental chambers that were used for the experiments. Although the material for all samples was CoPt, the methods of chemists and physicist differ very much. Additionally the essential difference between the two experiments is that one is for fundamental research and the other mainly for applied science. In experiments of fundamental research usually much energy is spent to determine all parameters of the experiment very precisely. This quite often raises the time investment for a single sample significantly and sometimes reduces the lifetime of samples. In applied science some methods to reduce the labor cost for a single sample lead to less well defined samples. On the other hand applications can only be realized if some way of mass production is possible.

3.1 Introducing wet chemical CoPt nanoparticles

Wet chemical processes have the general advantage with respect to evaporation or sputter processes, that the output is macroscopic. This makes samples prepared in wet chemical processes candidates for actual industrial usage. However one has to accept the usually less defined structure of the samples. One great challenge in wet chemical synthesis processes is to reproduce the properties of the particles, size and stoichiometric structure, in different batches. Particle sizes can be chosen in wet-chemical processes by stopping the reaction. One way to achieve this is by adding a chemical to stop the nucleation process. This procedure allows to produce particles in a wide size range. However it limits the accuracy of the size reproduction in different batches, since it is not possible to measure particle size during the synthesis. One advantage of the process is the freedom of choice of the organic ligands. One can use any ligand in the synthesis, as long as the ligands do not limit the size of the particles on there own. The second way to choose particle sizes in wet chemical processes is by choice of organic ligands that hinder the growth of the particles beyond a certain size. Hence, the sizes are easily reproducible. However the different size that can be produced are determined by the different organic ligands available for the specific synthesis. The sizes of the CoPt nanoparticles measured in this work were controlled in a third way. In the hot organometallic synthesis used one has the freedom of choice of the organic ligands, while still having a high reproducibility of particle size in different batches. The particle size is mainly controlled by the reaction temperature (figure 3.2), allowing to tailor CoPt particles from 1.5 nm up to 14 nm in diameter, which is a broad range for wet-chemically synthesized nanoparticles.

The measured samples were produced in the Group of Professor Weller of the institute of Physical Chemistry in Hamburg. In this specific hot organometallic synthesis platinum acetylacetonate is reduced while cobalt carbonyl is thermo-decomposed in a solvent containing 1-adamantanecarboxylic acid, hexadecylamine and diphenyl ether. For the synthesis a standard Schlenk line technique [2, 151] was used and thus the entire synthesis process, including the fresh preparation of cobalt stock solution was carried out under nitrogen atmosphere. The wet chemical particles under study process chemically reactive surfaces, which was used to undergo procedures as ligand exchange after synthesis [2] and it was therefore suspected that there could be at least a partial surface oxidation. To distinguish the oxidation that might occur already during the synthesis from the oxidation of aging and contact to air after the synthesis process, 3 different particle sizes (4 nm, 6 nm and 8 nm) were synthesized and prepared completely in a glove-box under nitrogen atmosphere. These samples (labeled M1, M2 and M3) were deposited by spin coating in the glove box under nitrogen atmosphere and stored under nitrogen until measured. After synthesis the solutions are washed several times, centrifuged and filtered. The bigger the synthesized particles were the fewer were the number of needed washing cycles. This could mean that there are more precursors left when synthesizing small particles, since the amount of cleaning cycles used is derived by the times of cleaning cycles needed, until the clusters grow together, minus two cycles. This should leave all clusters with a rather similar amount of not removed precursors per surface area of the particle. By this processing the size distribution of the particles can be narrowed down to 10%, which is a very sharp size distribution for chemically prepared particles. All post synthesis processes, as washing and film preparation were usually performed in air. The entire synthesis process is discussed much more detailed in the literature [123–125].

Keeping the particles in an oxygen-free environment during the entire synthesis and deposition process increases the amount of work for the chemists strongly, therefore one of the goals of the first beamtime was to investigate whether there were any differences noticeable between nanoparticles produced, deposited and stored until measurement in an oxygen-free environment and particles that were just synthesized that way. Despite the different preparations absorption measurements have shown no distinguishable differences. Hence in the two following beamtimes only particles that were only synthesized in an oxygen free environment but deposited and stored in contact with air were measured.

3.1.1 Sample preparation and characterization

After synthesis and washing procedures the nanoparticles were stored in a solvent (2-propanol, chloroform or as for the here measured samples: toluene), the solutions were diluted for the coating processes. The depletion procedure is not as precise as the size selectivity, since the yield of the synthesis process varies between two batches. This way the prepared samples ended up with coverage down to 1/3 of a layer and up to 3 layers. A p-doped (Bor) Silicon Wafer with a resistance of 1-20 Ohm/cm was cut in suitable pieces of 8

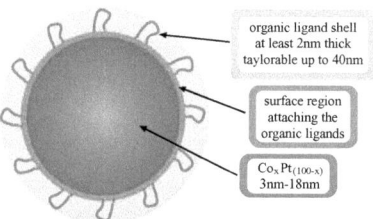

Figure 3.1: The basic structure of the particles is a core of Co and Pt and a shell of organic ligands, that are attached to surface of the Co-Pt-Core. The Core is most probably structured in face centered cubic with statistically occupied lattice sites. The changing stoichiometric ratio with particle size is known, but not precisely studied at this point.

· 8 mm² (figure 3.5). Acetone and isopropanol were used to clean these pieces. We chose to leave the natural oxide on top of the silicon wafer, instead of removing it as it is often done (with $HF_{(aq)}$), because the conductivity of the samples was high enough, that no charging took place during the SEM characterization prior to the XPS measurements. Above that there is no danger due to the organic ligand shell surrounding the nanoparticles, that oxygen out of siliconoxide bonds may be transferred to oxidize the nanoparticles. The silicon wafers were coated with the nanoparticles using three different techniques: By spin-coating, dip-coating or means of Langmuir Blodget (LB) technique. Spin coating is the method of choice if one tries to vary the coverage in a broad range independent of the ligands attached to the particles, but is different to the other two methods rather impracticable for industrial processes. Dip coating and the LB technique have requirements to the solvents and the ligands in order to work properly. LB is the most valuable technique to prepare molecular monolayers. Dip coating is the sim-

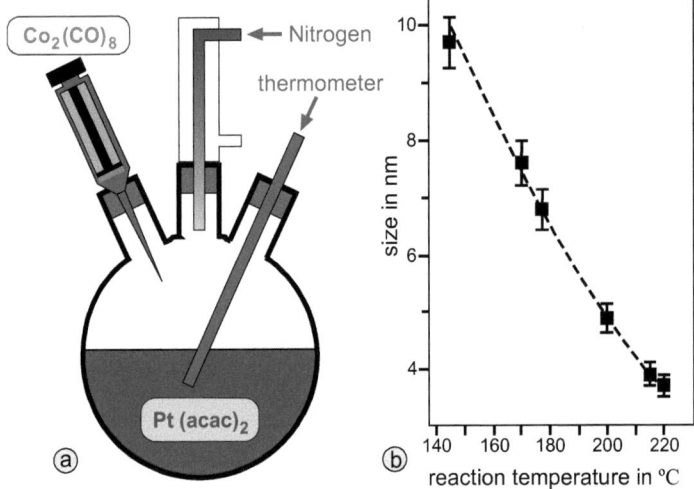

Figure 3.2: (a)The CoPt nanoparticle synthesis is done under N_2 atmosphere. Platinum acetylate is heated in solvents to a certain temperature, when this is reached the cobalt precursors are injected. (b) The reaction temperature is crucially affecting the size of the synthesized nanoparticles (figure from reference [123]). By varying the synthesis temperature, while keeping all other synthesis parameters constant, the resulting CoPt nanoparticles vary in size, growing larger at lower reaction temperatures, thus opening a size range to tailor the particles from 3 nm to 18 nm (the graph includes only data points up to 10 nm sized particles).

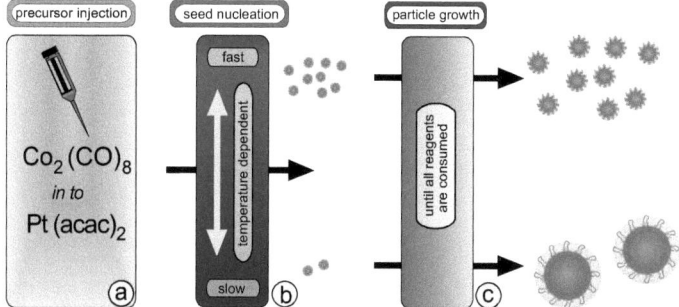

Figure 3.3: The growing of nanoparticles: During the Synthesis, the cobalt precursors form nucleation seed which then grow to CoPt nanoparticles. The seeding speed varies significantly with the reaction temperature, leading to many seeds at higher and rather few seeds at lower synthesis temperature. The growing process on the other hand is more or less independent of the temperature, hence the particles are more, but smaller at higher synthesis temperatures.

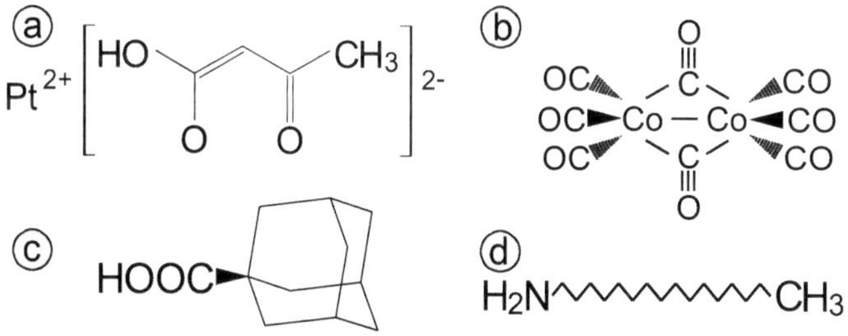

Figure 3.4: Structural formula of the important chemicals used in the CoPt nanoparticle synthesis: (a)platinum-acetylacetonate, (b)cobaltcarbonyl, (c)1-adamantan-carboxylic acid (ACA), (d)1-hexadecylamin (HDA)

plest technique, with the least amount of technical setup and thus the preferred industrial application. All prepared films were analyzed using SEM. We chose solution concentrations and preparation settings to result in coverages closest to one layer of nanoparticles. Some especially prepared Carbon grid samples had already been measured by transmission electron microscopy (TEM), showing the nanoparticles to be very homogeneous. A possible deformation after deposition (figure 3.6), as it has been shown for 12 nm FeCo nanoparticles by M.Getzlaff et al [51] can not be excluded.

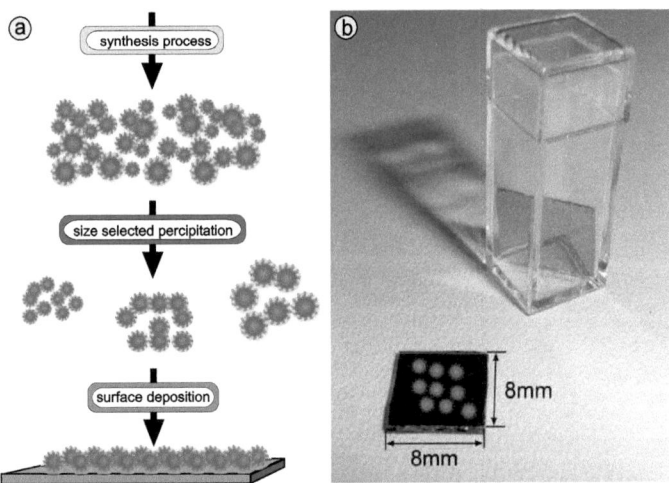

Figure 3.5: (a) During the synthesis different particle sizes are created with a pronounced broad maximum at a certain size, which can be selected by the right synthesis temperature. The size distribution can be sharpened by size selective precipitation. Finally the particles are deposited on a surface (in this case silicon wafer) by standard deposition techniques as spin-coating, dip-coating or the langmuir-blodget-technique. (b) The final samples were in this measurement silicon wafer pieces of $8 \cdot 8$ mm^2.

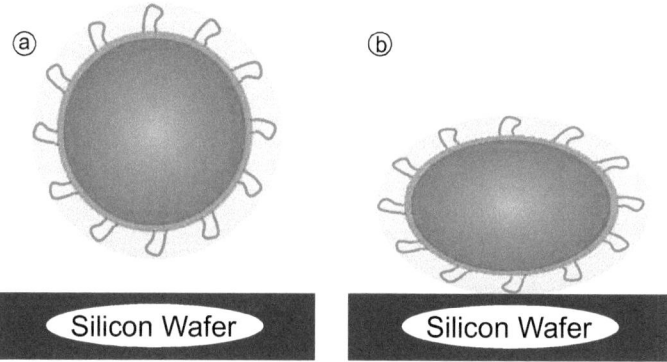

Figure 3.6: Particles deposited upon a surface might change their appearance from almost circular shape (a) to a rather lenticular (b) like shape and thus change their magnetic properties

The experimental chamber at the beamline was equipped with a load lock system to change the samples and with a Helium cooled super-conducting high field magnet supporting a field of up to 7 Tesla during the first beamtime and up to 6 Tesla during the latter two [1]. The magnetization curve in figure 3.7 indicates that a field of at least 2 Tesla is needed to magnetize the nanoparticles to a high extend (above 95%). The chamber was setup properly to fit the sample in the focus of the beamline which is 100 μm· 60 μm.

The exit slit was set to typically 100 μm, thus allowing an energy resolution of roughly 200 meV in the energy range 750 eV to 850 eV. The Spectra were taken with 200 meV steps between two neighboring data points. While measuring the absorption spectra, the

---

1. unfortunately the magnet had been quenched between our first and second beamtime several times. When trying to change the magnetic field of the superconducting magnet too fast, the increasing current may temporarily destroy the superconductivity and then completely dissipate the electric energy of the into heat. The process is called quenching and permanently damages the superconducting magnet, reducing the maximum field it can sustain.

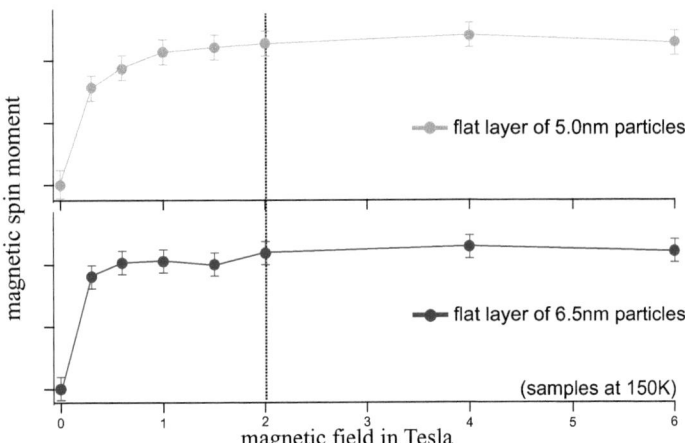

Figure 3.7: Increasing the magnetic field, while measuring the same sample shows a slight different magnetic spin saturation behavior for different size nanoparticles. An almost complete saturation reached at 2 Tesla applied external magnetic field.

helicity of the light was switched, rather than the magnetization direction of the magnetic field, since both measures have the same effect on the absorption behavior of the sample and switching the helicity was much faster with this chamber than switching the magnetic field. The direction of the magnetic field was always parallel to the incoming photons from the beamline. Most of the spectra were taken in normal incidence of the light to the sample surface, while a few were taken varying the angle of the sample surface up to 60° out of the normal geometry (figure 3.17) , leaving the magnetic field parallel to the incoming light.

3.2 Mass selected Co_nPt_m clusters

The fundamental research experiment of mass selected small Co_nPt_m clusters deposited on magnetic substrates, highly depend on the experiment ICARUS. The Name ICARUS is a relict from early sputter experiments (Ionic Clusters by ARgon spUttering Source), where the sputtering was tested with argon, today the heavier more efficient sputter gas xenon is used. ICARUS is a travel Experiment, designed to be quickly disassembled and reassembled with as little calibration work as possible. The experiment consists of three parts, the sputtering source for cluster, the dipole magnet for mass selection and the spectroscopy chamber.

3.2.1 Sputter chamber and mass selection

The sputter source is a UHV chamber with an operational base pressure of $5 \cdot 10^{-8}$ mbar. The cluster source is differentially pumped with a 550 l/s, two 240 l/s and a 70 l/s turbo molecular pumps, with a M15 Eco Dry as roughing pump (now replaced by a scroll pump). As in many sputter sources the use of relatively high pressured noble gases for the sputter process makes this differentially pumping and a high pump rate necessary. This unique sputter source uses 30 kV to accelerate xenon ions to sputter any solid (not insulating) target between 1 cm and 3 cm in diameter (figure 3.8).

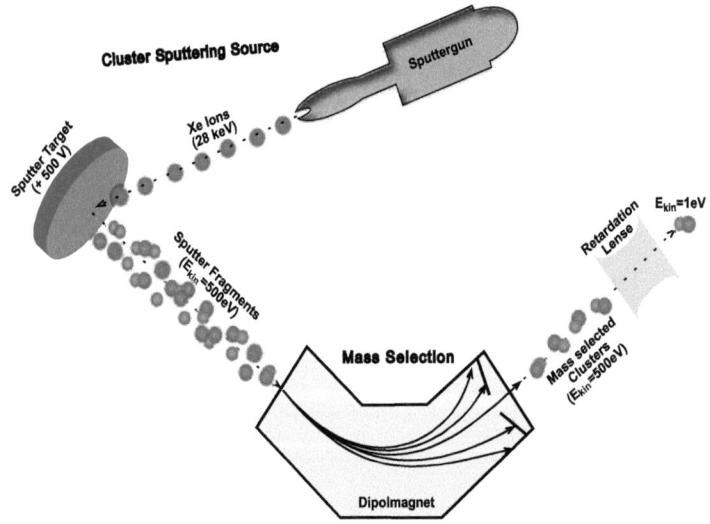

Figure 3.8: 30 keV xenon ions are focused on a sputter target, which is kept on a constant potential of +500 V. The positively ionized sputter fragments are repelled by the positively charged sputter target and accelerated towards a dipole-magnet. Mass selection is achieved through the dipole magnet and the cluster beam may be decelerated by a retarding lens to below 1 eV per cluster.

The sputter target is set to a potential of usually U_{acc}=+500 V to accelerate the positively charged clusters away from the target, forming an unfocused ion beam. This ion beam is focused by electrostatic ion lenses system that is powered by CAEN Sy2527 multi power supply. For mass separation the cluster beam passes a double focusing sector magnet that separates the fragments by $\frac{mass}{charge}$. Charged particles ($q \cdot e$) in a magnetic field will move according to their kinetic energy and their mass on radial paths (formula 3.1).

$$\frac{m}{q \cdot e} = \frac{e \cdot B^2 \cdot r^2}{2 \cdot E_{kin}} = k \cdot B^2 \tag{3.1}$$

In this case the radius of the magnet is fixed, the kinetic energy ($U_{acc} \cdot e$) is the same for all clusters and there are very few doubly or multiple charged clusters. Please mark that in this work as it is common, all shown mass spectra are displayed as cluster signal against mass. The actual mass separation process samples by mass over charge, therefore it would be correct to label the axis as mass/charge. Most sputtered clusters are neutral [131] with decreasing probability with increasing charge. The amount of doubly and triply charged monomers can be seen in the mass spectra in the region below the singly charged monomer peak. For the CoPt mass spectra the multiply charged Co monomers were below of 5% of the singly charged monomer. The Co_1 peak in figure 3.9 is cut off for that figure, it reached 12.5 nA of maximal cluster current. There are many uncut mass spectra displayed in the mass spectra section of the Appendix A.

The applied magnetic field in the magnet allows to mass separate up to $\frac{\Delta m}{m} \leq 2\%$ (typically slightly above 3%). This allows to still separate Cu and Ni monomers from a CuNi-alloy target. Tuning the field from 0 T up to 0.55 T while detecting the selected cluster current, mass spectra of the used sputter targets can be recorded. In this manner several mass spectra of various targets were recorded, as the CoPt mass spectra shown in figure 3.9. A detailed description of this cluster source is given in the PhD Thesis of H.-U.Ehrke [37] and later by J.T.Lau et al. [78].

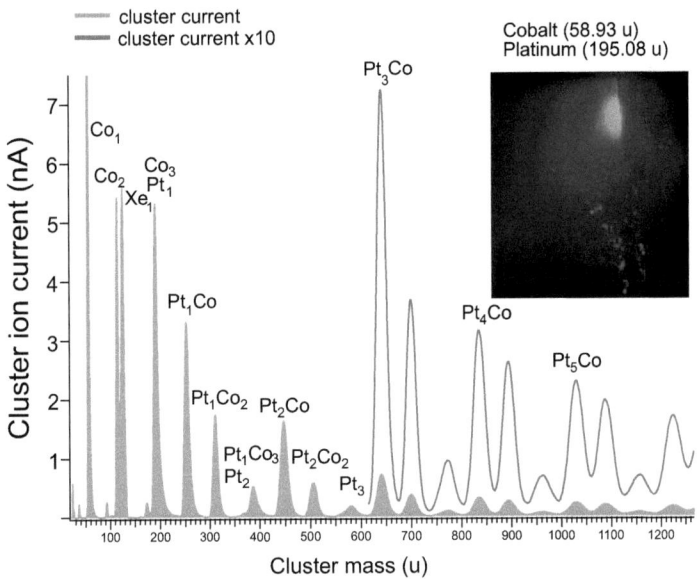

Figure 3.9: The size distribution of a a $Co_{25}Pt_{75}$ alloy target: Tuning the magnetic field of the dipole magnet from zero up to 0.5 T it is possible to record mass spectra of the sputter target. In this case it is a mass spectrum of a $Co_{25}Pt_{75}$ alloy target and the cluster beam was not retarded, which is obvious due to the presence of the xenon signal of low energetic xenon ions, which will completely disappear when the cluster beam is retarded with the sputter target potential. The Co_1 peak was cut off in intensity for this figure it's actual corresponding cluster current is 12 nA. The inset shows the optical appearance of the sputter spot, where the xenon ions impact (figure 3.8).

The cluster beam can be refocused to a small beam diameter of less than 1mm radius, which is done when measuring mass spectra or depositing clusters on surfaces. For focusing of the beam and mass spectra the current at the electrically isolated pin (2 mm diameter) of the manipulator is used (figure 3.12b). The actual spot geometry of one deposited cluster preparation was mapped by tuning the beamline to the L_3 edge absorption maximum of the cluster material (in this case cobalt) and taking the tey signal at different positions. The measurement shows that this spot had a diameter of roughly 1 mm and a gaussian like shape (figure 3.10).

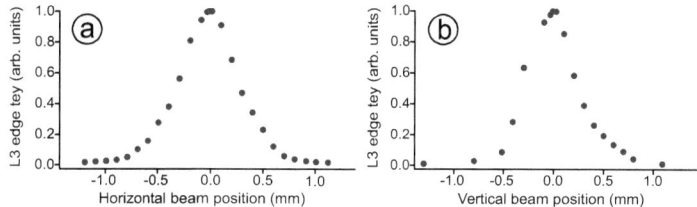

Figure 3.10: The cluster spot was measured with the L_3-edge tey and as the spatial distribution shows it is about 1 mm in diameter.

The deposition time t to establish a coverage of about 3% of a layer of clusters in the center of the deposited spot is calculated as follows. The supporting material is in most measurements a copper crystal with a 3 ML iron thin film that grows pseudomorphic in this regime. One can assume that the deposited 3% of a layer clusters grow pseudomorphic too. There are about $1.3 \cdot 10^{13}$ copper surface atoms/mm². The mass selection in the dipole magnet allow only singly charged clusters of the desired species to pass. One can now calculate the amount of clusters needed to cover the area of the cluster spot with 3% of a ML and from this charge estimate the deposition time depending on the cluster current, the cluster size and the area of the cluster spot:

$$t(s) = \frac{65s}{I_n(nA) \cdot n(\text{atoms per cluster}) \cdot A(mm^2)} \qquad (3.2)$$

Using other crystals or systems that may not be assumed to grow pseudomorphic in the lateral direction, the time would have to be changed according to the surface area the average deposited cluster would cover compared with the copper lattice used in this calculation. Due to the fixed size of the pin the cluster beam being focused upon one can only state prior to the experiment that the focused cluster spot on the sample will be equal or smaller than the dimension of the pin. It is impossible to scan the dimensions of the cluster spot at every measurement, hence it is typically assumed that outer perimeter of the cluster spot is equal to the that of the pin. This assumption proves true in the light of the data shown in figure 3.10. The formula 3.2 implies a constant cluster current during the entire time of the deposition process, which is not possible with the cluster source at the moment, since the sputter spot stays localized at one position on the sputter target. That way the sputter target is eroded while depositing clusters and due to the deepening of the hole the cluster current decreases gradually. While depositing cluster sizes with higher yields the deposition time is often under 5 minutes and the cluster current decreases less than 10%, but when depositing 15 minutes the cluster current can drop down to 50% of it's initial value. It is thus advisable to monitor the cluster current during the deposition cycle and take it into account.

3.2.2 Sectroscopy chamber

The spectroscopy experiments were performed in a ultra high vacuum chamber with a base pressure of $2 \cdot 10^{-10}$ mbar. This is crucial to assure sample lifetimes of several hours, which are necessary to record the data of a typical cluster preparation. The time for residual gases to completely cover a clean surface in a $1 \cdot 10^{-6}$ mbar surrounding is about a second, at $2 \cdot 10^{-10}$ mbar the process takes roughly 2 hours, which in case of highly reactive clusters may even be too fast. For this reason the spectroscopy chamber that at first had been used for film preparation and cluster spectroscopy was redesigned and split into a preparation chamber and a spectroscopy chamber, that are connected by a valve(figure 3.11). The new spectroscopy chamber has a base pressure of $8 \cdot 10^{-11}$ mbar, extending the lifetime of the samples by a factor of 2.5. The spectroscopy chamber is pumped by a 550 l/s turbo molecular pump, a nitrogen cooled titanium sublimation pump (TSP) and an ion getter pump, which can be powered by batteries and thus allowing to transport the chamber under high vacuum. There is a SES-100 Electron analyzer and a Al and Mg K-α x-ray gun mounted. A Leybold Ionivac IM540 ion gauge is used to measure the pressure. The beamline access is equipped with a differentially pumped tube with a thin opening to avoid contaminating the chamber with residual gases from the beamline of the storage ring. This tube is pumped by two cascaded 70 l/s turbo molecular pumps and a membrane roughing pump.

The preparation chamber is used for the sample cleaning and in situ preparation of the samples.It is pumped by a 550 l/s turbo molecular

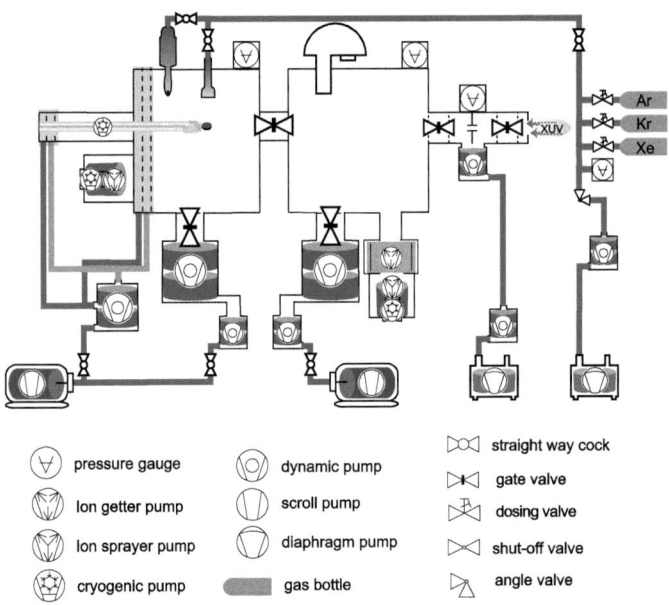

Figure 3.11: Here is an overview of the ICARUS preparation and spectroscopy chamber with all vacuum components typically attached.

pump and a nitrogen cooled TSP, additionally the revolving front flange and the manipulator are each differentially pumped with a 2 staged Leybold Turbovac 151. Network cameras installed allow visual control of the inside, making the chamber manageable by only one scientist. The chamber is equipped with the retarding lens to decelerate the mass selected clusters, an argon sputter gun, a LEED optic, three metal foil evaporators, a coil to magnetize thin films, a gas doser, a mass spectrometer and a manipulator. The manipulator is equipped with a newly constructed two sided head (figure 3.12). One side is used for measuring the mass spectra and for cluster beam tuning, while the other side is holding a Cu(100) single crystal (0.5° accuracy) for the spectroscopy experiments. The surface of the Cu crystal and the pin of the backside have the same distance to the revolving axis to make comparable measurements on the two opposing sides possible by just revolving the manipulator 180°. The crystal side is equipped with a heating circuit, that allows radiant and electron bombarding heating. The manipulator can be cooled by liquid nitrogen or liquid helium, allowing to reach crystal temperatures as low as 35 K [2]. A temperature 35 K is sufficient to freeze out argon and all heavier noble gases, as required for the soft landing process used in the sample preparation. Temperatures are measured with chromel alumel thermocouples one attached to the crystal and one to the tip of the manipulator and in case of low temperatures cross checked by noble gas thermal desorption spectroscopy (TDS).

---

[2]. The crystal is mounted by a laser welded connection to an iridium rod. If this rod is not sufficiently tempered in advance, it's thermal conductivity at low temperatures may hinder the crystal to cool to temperatures below 45 K.

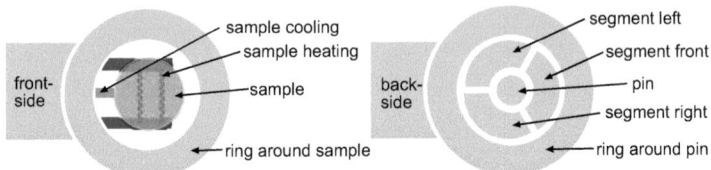

Figure 3.12: In this graph the two sides of the manipulator head piece is pictured. The backside is used to localize the cluster beam position in space and to measure mass spectra, while the front side contains the liquid helium coolable Cu(100) single crystal and a heating system, which is needed for sample preparation and the spectroscopic measurements.

The argon sputter gun and the gas shower are connected to a gas dosing system (baked out before usage), which consists of three leak valves with attached lab gas bottles and a Baratron pressure gauge. It is pumped by a 70 l/s turbo molecular pump attached to a membrane pump. The gas pressure can be adjusted from $1 \cdot 10^{-4}$ mbar up to $1 \cdot 10^{-1}$ mbar which allows maximum argon sputter yields in the cleaning process at $3.5 \cdot 10^{-3}$ mbar pressure and the dosing of single complete layer of xenon to the crystal at $4.0 \cdot 10^{-4}$ mbar in 3 seconds.

The three newly constructed foil evaporators have a water cooling within both power connectors, which replaces the former one sided water cooling. This improved the time stability of the evaporators, reducing the drift due to heat load, keeping it completely stable during the 10 min evaporation cycles.

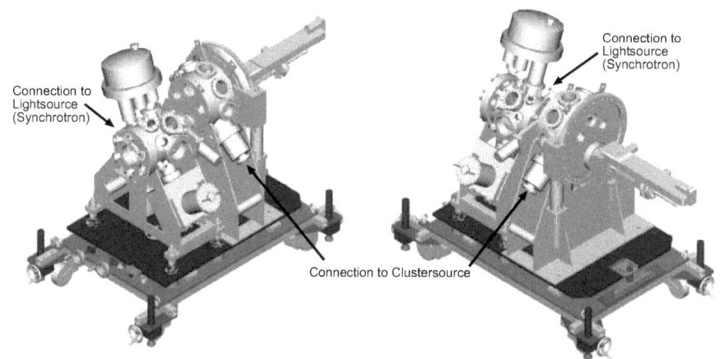

Figure 3.13: The new ICARUS spectroscopy chamber

3.2.3 In situ sample preparation

Samples that consist to a great extend out of surfaces usually age quickly, due to the high contact to residual gases surrounding it and due to a typically strongly enhanced reactivity compared to bulk materials. This is true for all clusters prepared with ICARUS so far, that way it is only possible to measure in situ prepared samples. The systems measured are magnetic clusters in contact with magnetized surfaces. The exchange interaction to the magnetic surface aligns and magnetizes the deposited clusters. The geometry of our chamber allows normal to gracing incidence of the light onto the crystal. To maximize the XMCD signal measured we prefer layered systems with a possible perpendicular magnetic anisotropy, such as Fe on Ru(0001), Fe on Cu(100) and Ni on Cu(100).

The preparation cycle of iron on copper will be discussed as a typical example as depicted in figure 3.14. The crystal is sputtered with the argon sputter gun for 15 minutes, while the ion current is monitored to control the process, adjusting the argon pressure in the gas dosing

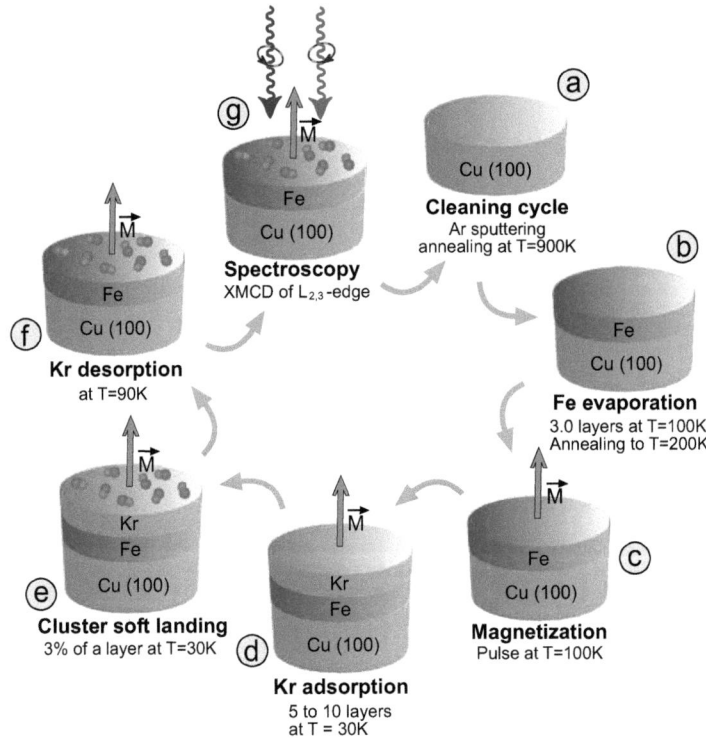

Figure 3.14: A typical in situ preparation cycle includes a sputter anneal process (a) of the copper crystal, the preparation of a thin film (b) and its magnetization by a field pulse (c). After sufficient cooling 5-10 layers noble gas matrix is frozen onto the thin film (d), followed by the loft landing of 3% of a layer of clusters (e), after which the noble gas matrix is carefully thermo-desorbed (f). Finally the spectroscopic measurements at the deposited clusters can be performed (g).

system keeping the sputter current at 1 to 3 μA. The sputtered Cu(100) single crystal will be annealed 1 minute to 900 K. At a crystal temperature of less than 200 K, the 3 layers of iron will be evaporated within 5 minutes onto the crystal using a foil evaporator with a 100 μm pure iron foil, applying 800 mV to the evaporator. The calibration of the evaporators was done with TDS and has been described with the method 2.1.2. This film will be flash-heated to 200 K and magnetized when the temperatures drops below 100 K, applying 2-3 short magnetic pulses perpendicular to the surface. The Fe/Cu system is perpendicular magnetized and will stay that way four over 12 hour, even when reaching room temperature as a test has shown. After the crystal has been cooled to temperatures below 50 K, 6 to 7 layers of krypton will be frozen out on the crystal surface, using the gas shower for 60 seconds, keeping $8.2 \cdot 10^{-4}$ mbar krypton gas pressure in the gas dosing system. The thickness was always verified in the later desorption, using the mass spectrometer for TDS. Into this Noble gas cushion the clusters are now soft landed with a maximum of 1 eV/cluster kinetic energy. Soft landing is done to prevent the clusters from segregation upon impact (see detailed discussion for this cluster source by J.T Lau [79]). The time of the deposition process depends on the intensity of the cluster beam and is evaluated using formula 3.2. The preparation cycle ends with the careful evaporation of the frozen noble gas, allowing the cluster to slowly settle onto the surface. During the noble gas matrix evaporation the ratio of monolayer to multi-layer peak of the evaporating noble gas is measured with TDS.

3.3 Data acquisition

Most of the measured data of this work was acquired in several short measuring periods at the storage ring center BESSY II in Berlin. The typical time periods of measurement was 1 week for the wet chemically prepared nanoparticles and 2 weeks of beam plus one week of set up time for the mass selected cluster experiment. All experiments were done at the soft x-ray edges of cobalt, iron and nickel and thus had to be performed in UHV conditions, since the absorption length of soft x-ray light is only a few cm in air and thus would make any measurements in air impossible. Due to the limited actual measuring time at the storage ring, it is highly necessary for the success of any experiment that everything runs perfectly in this short time period.

The chemically prepared particles were all prepared as samples in advance, additionally the chamber used belonged to a French cooperation group and was all set up and running, when starting the measurements. The mass selected clusters were prepared in situ and with the own experiment, which is quite large for a mobile experiment and thus the system had to be set up at the storage ring in advance to the measurements.

In this section there will be a short introduction to the storage ring and the beamlines used, than the general setup at the storage ring will be explained, how data is acquired and the standard procedure to remove the background.

3.3.1 Experiments at a storage ring

The absorption measurements were carried out in Berlin at the storage ring center Bessy II, using the soft x-ray beamlines UE46, UE52/SGM and UE56/2. These beamlines can emit circularly polarized light in the energy region of the transition metal L-edges using the third harmonics of the undulator providing a degree of polarization of 90% (table 3.1).

The light is produced by relativistic electrons that circulate the storage ring and are forced on oscillatory paths perpendicular to their translational motion, thus emitting light mainly straight forward. A brief but nice introduction about synchrotron radiation can be found in the articles of G.Margaritondo [93, 94]. The devices used to initiate the changes in the electron path can be bending magnets, wiggler or undulators. Bending magnets change the path of the electron only once but very strong, which leads to a short pulse and a broad bandwidth of the emitted light. A wiggler consists of a series of alternating bending magnets, thus changing the path of the electron quite strong and leading to a series of broad band-

Beamline	UE52/SGM	UE46 PGM/1	UE56/2 PGM/2
energy region (eV)	90-1500	130-1600	60-1300
circular polarization (%)	90	90	90
photons @750eV (I_{sr}=100 mA)	$1 \cdot 10^{11}$	$2 \cdot 10^{12}$	$2 \cdot 10^{11}$
slit setting (μm)	100	100	100
energy resolution @750eV (meV)	150	150	250

Table 3.1: Some specifications of the beamlines used for the measurements in this work

width light pulses. An undulator evokes relatively slight changes of the electron path, leaving the electron main direction towards the destination of the light (experimental setup) and by that creates a continues pulse of light with a narrow bandwidth. All the beamlines used for the measurements in this work were undulator beamlines [146], a typical layout can be seen in figure 3.15.

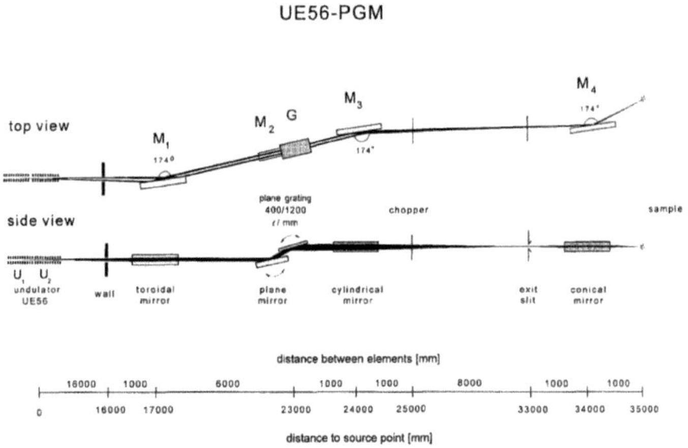

Figure 3.15: A typical beamline: It consists (from left to right) of an insertion device (undulator) producing light with a certain bandwidth, then a mirror to focus the light onto the monochromator, after which two mirrors and an exit slit are used to focus the light at the location of the experiment. This layout of the UE56-PGM beamline at BESSY II ist taken from reference [115]

To be able to narrow down the bandwidth of the light even further all of the used beamlines were equipped with monochromators. The most common at BESSY is the SX700 [107] that uses dispersion at plane gratings to be able to select a specific wavelength of the light to pass through an exit slit. Finally the beam is refocussed again in horizontal and vertical direction to illuminate a well defined region of the experiment, depending on the desired focus, typically between 20·20 μm^2 to 100·100 μm^2.

The experiments were set up at the end of the beamline, as it is shown in figure 3.16 for the high field chamber, that was used to measure the wetchemically prepared nanoparticles.

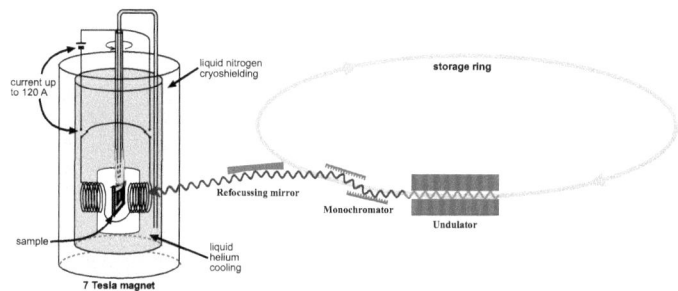

Figure 3.16: Scematics of the high magnetic field chamber at the Beamline at Bessy II

If the user desires to measure rather a larger area of the sample to integrate over more particles, the experiment may be placed out of the focus of the beamline. Depending on the effect that are to be measured the sample can be placed in different geometries to the light. Most of the measurements in this work were done in normal geometry (figure 3.17), which means that the light shines normal onto the surface of the sample. For the wetchemically prepared nanoparticles the geometry was changed for some measurements, but for all mass selected preparations only normal incidence was used.

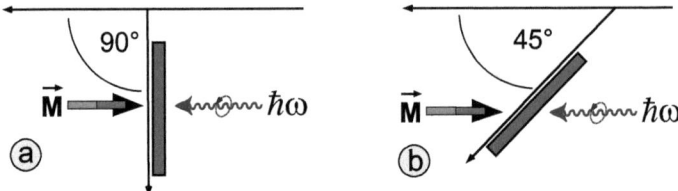

Figure 3.17: Most samples were measured in normal incidence (a) of the light from the synchrotron. In some cases the angle of the surface normal and the light was changed (b) up to 60%, while light and magnetic field remained parallel

3.3.2 Tey measurement

All XAS measurements in this work were done measuring the tey. There were slight differences in the technical realization of the recording of the signal, since two very different experiments were used. The physics was identical, hence the description of the data acquisition is done for the actual setup used at the mass selected clusters. For measuring the tey the manipulator is grounded except for the sample. The current of the sample is fed into a current to frequency converter. The output of which is sent to a counter card (figure 3.18).

To normalize the measured signal to features induced by the beamline and monitor the general status of the synchrotron several additional channels were recorded: The current of the last refocusing mirror of the beamline to normalize the sample signal to the incoming photon flux[3]. The current of the ion gauge in the measuring

---

[3]. This can similarly be done with the current of a gold mesh inserted into the beam, but these meshes are quite often contaminated with all kinds of unknown adsorbates. Those adsorbates not evenly distributed over the gold mesh, which leads to distinct changes in the absorption signal if the x-ray beam slightly drifts in its position. Additionally the adsorbates may desorb or fragment during the measurement changing their absorbtion characteristic, which

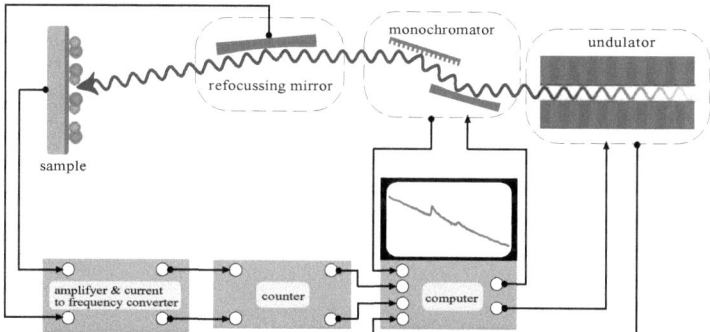

Figure 3.18: A typical setup for tey measurements. A computer is used to control undulator and monochromator, to select the photon energy, while the current of sample and refocusing mirror is recorded directly or as in this displayed case with the help of a current to frequency converter and a counter card

chamber[4], the position of the monochromator, the shift and gap of the undulator, as well as current and lifetime of the electrons in the storage ring. The measuring program was the BESSY in house measuring program EMP 2. EMP 2 allows to define x-files of the energy positions, which one wants to measure at. This way we were able to measure with different energy resolutions depending on the position within the measured spectrum. For the measurements x-files were used (table 3.19). This way one can save precious time, by reducing the resolution in regions with little to no change in the spectrum and keep a high resolution at the absorption edges. During a typical sample measurement the magnetization of the sam-

---

if the gold mesh signal is used in the normalization process may leave an undesired imprint on the data set. If the gold mesh is equipped with an gold evaporator (which is not the case in the beamlines used), those effects on the gold mesh signal can be removed by burying the adsorbates on the mesh under fresh layers of gold.

4. this was not be recorded in the high field chamber, since an ion gauge does not display any reasonable values with a high magnetic field presence

energy (eV)	step (eV)
755 - 768	1.0
768 - 803	0.2
803 - 820	1.0

energy (eV)	step (eV)
680 - 700	1.0
700 - 736	0.4
736 - 750	1.0

Figure 3.19: The x-files used for the data acquisition were created to produce energy resolution for cobalt (left table) and iron (right table) with a higher resolution at the absorption edges.

ple was verified. The high field magnet displayed the current and the magnetic field present. In the case of the mass selected clusters magnetization of the substrate (iron or nickel) was measured taking a left and right circular scan of the $L_{2,3}$-edges. Only if the substrate XMCD signal was sufficiently strong, the cluster signal was recorded measuring 4 times changing the helicity of the light twice, thus recording the first and last spectrum with the same helicity. The signal from the wetchemically prepared nanoparticles was very strong and thus only one set was recorded for each magnetic field chosen, except for the highest, were the recording was performed as in the case of the mass selected clusters (+,-,-,+ or -,+,+,-). In both experiments the magnetic field was kept constant and the helicity of the light was changed, although that usually influences the transmission of the beamline and the position of the spot slightly and thus causes some small uncorrectable errors in the data. On the other hand the magnetization reversal may cause stronger problems, if the field is slightly different. In the case of the two experiments the question was rather academic, since the high field magnet would have needed roughly 30 Minutes to invert the magnetization (compared with 1 Minute for the undulator). The substrate of the mass selected clusters on the other hand can not be magnetized at the measuring position, thus would have to be moved and this would make it impossible to measure the exact same position twice, making any measured data completely useless.

3.3.3 Tey background treatment

The original measured spectra have several features folded into the signal: i) The gradually diminishing transmission of the beamline towards higher photon energies. ii) Presently unavoidable at BESSY II is the continuously decreasing ring current of the storage ring[5], which evokes a falling background. The unwanted signal components can be removed to a large extent by normalizing the signal to the current measured at the last refocussing mirror.

In case of the CoPt nanoparticles the tey signal is strong (signal to background ratio 2:1), comparable with a several layer thick thin film, that support the mass selected clusters. After normalization to the last mirror current remains only a slightly falling background that represents the background due to the silicon wafer (respectably the copper crystal) and will be fitted by a straight line and then removed. The proper treatment for it is the linear fitting in the pre-L_3-edge region of the spectrum and subtraction of this linear function from the entire spectrum. All processing steps are illustrated in figure 3.20.

The magnetic properties of the $3d$ elements are almost entirely due

---

5. This effect is due to the injections of new electrons into the storage ring every 8h. The most recently build storage rings use the so called top up mode, which allows an almost continuos injection of new charge carriers. The injection of new particles every minutes is much more demanding, but if it is achieved allows the beam to be used at a constant intensity and almost completely removes the heat load fluctuations within the beamline due to ring current changes.

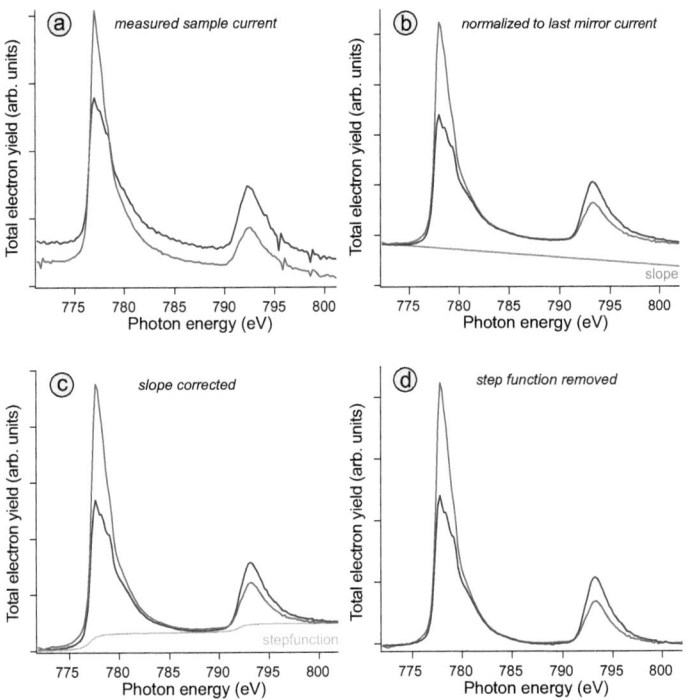

Figure 3.20: An illustration of the main data evaluation steps: The total electron yield raw data measured at the Co $L_{2,3}$-edges (a) has to be normalized to the photon flux, using the last mirror current, then removing the slight slope (b) of the falling background and the step function (c) originating in non $2p$ to $3d$ excitations, one receives the pure resonant $2p$ to $3d$ absorption signal (d).

to the not completely filled d-states. For the application of the XMCD sum rules (formula 2.3 and 2.4), which only consider excitations from $2p$ to $3d$ states, the spectrum has to be stripped of all $2p$ to continuum and $2p$ to s-state contributions (section 2.1). This is typically done by subtracting a hyperbolic step function with two steps, one at the L_3 and one at the L_2-edge with a step height ratio of two to one (section 2.1.1) resulting in a spectrum consisting only of the contributions from $2p_{1/2}$ and $2p_{3/2}$ to empty d-state transitions. All spectra taken for the mass selected cluster substrates and the wetchemically prepared nanoparticles have been equally treated in described way. Errors assumed for the sum rules are typically 10% [25].

The coverage of the mass selected clusters being only 3% of an atomic layer on the surface gives a very weak signal, which makes the background treatment slightly more challenging. First of all it is crucial to measure background. A spectra, that records the substrate features in the region of the deposited clusters, without the clusters present (figure 3.21).

Obviously the two backgrounds are very different for the nickel and the iron thin film. First of all there are absorption features at the Co L-edges in case of the nickel thin film and none when using the iron layer. Between the last beamtime with iron thin film as substrate layer and the beamtime with nickel as substrate, the switching mirror unit was plasma cleaned to remove carbon from the mirrors. The antenna to produce the high frequency electromagnetic field needed to ignite the plasma had a brass thread. The plasma cleaned off the carbon, but some of the brass remained in the vacuum chamber, producing new absorption features in the x-ray beam emitted by

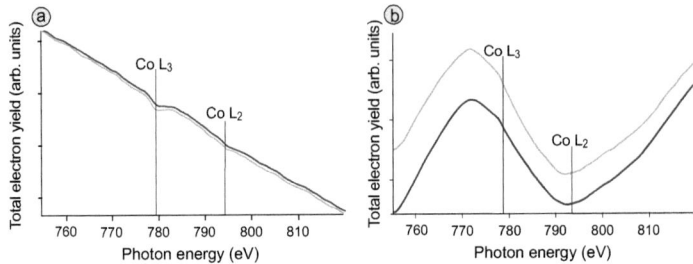

Figure 3.21: The background for the mass selected clusters is recorded in the region of the cobalt L-edges, for a nickel substrate (a) and an iron substrate (b). Very evident is the different background structure, while in (a) there are absorption features only visible at the Co L-edges (a beamline contamination) in (b) the beamline is clean, but there are strong features due to the iron substrate

that beamline. One absorption feature was at the cobalt L-edges. Otherwise the nickel background looks rather smooth, but not so for iron. The strong wiggling features there are iron EXAFS oscillation from the iron L-edges at $h \cdot \nu = 706.8$ eV and 719.9 eV.

The background spectra taken will be used to be subtracted from the measured cluster spectrum, if the background is really smooth. If the background has structure, it is better to divide the cluster spectrum by the background. This procedure has thoroughly been discussed by J.Stöhr [132] and successfully been applied for deposited mass selected clusters by J.T.Lau [77] and M.Reif [111].

The necessity to normalize the measured spectrum to the current of the last mirror to eliminate some effects (including contaminations of the last refocusing mirror) become more apparent, when measuring very small coverage of particles. Figure 3.22 displays on the left side the actual data measured for mass selected clusters and on the right, the signal after normalizing to the last mirror current. The

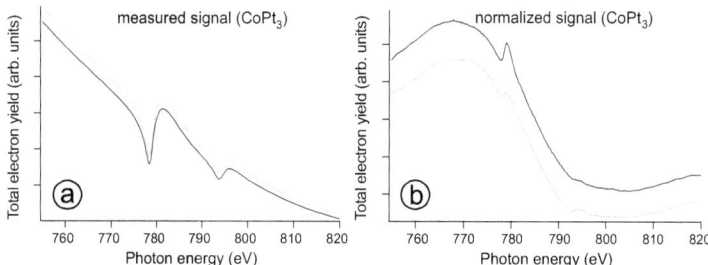

Figure 3.22: The raw data has to be normalized to the photon flux using the last mirror current. In the case of the deposited clusters, the necessity of this procedure becomes more obvious than in the earlier example of the wetchemically synthesized nanoparticles (figure 3.20a+b). Without the normalization the recorded data is completely dominated by the characteristic of the beamline.

basic procedure is the same for both experiments, only with the additional step of the division by the measured background, that follows right after the normalization to the last mirror current.

The treatment for the background of mass selected deposited clusters was performed as described above for all clusters deposited on nickel. In the case of the iron substrate a slight variation had to be used that will be discussed in the section of the results of the mass selected cluster measurements (section 5.1).

Chapter 4

Wet Chemical Nanoparticles

In this chapter the measured data of the wet chemically prepared Co_xPt_{100-x} nanoparticles will be presented and discussed. First the line shape and integrated areas of the white line spectra will be analyzed to estimate the oxidation state and the branching ratio (BR). Then the circular scans and the dichroism spectra will be investigated and the thereby derived magnetic properties will be presented.

4.1 Oxidation effects

As it is vital for any potential application one of the first questions when dealing with wet chemical methods is: How clean is the synthesis and did it produce the desired product. From the Chemist point of view the starting reagents can be measured, the left over reagents can be determent and the product can be measured, but whether some additional input was made from solvents or the atmosphere (air or gas) can not easily be answered. Measuring the XAS of Co it is very easy to distinguish between the spectrum of metallic Co from that of CoO and both from other Co oxides, since the oxidation state of the Co strongly influences the fine structure of the absorption lines. The first questions to answer are how clean are the samples leaving the synthesis and is there an aging effect when stored in contact to air? For the first investigation two sets of samples of equally sized nanoparticles were synthesized, one where the entire process of synthesis, preparation and storage was under nitrogen atmosphere and a reference set, where after synthesis the particles had contact to air until the measurement. To investigate the time dependent behavior samples of different age were investigated. Comparing sets of similar and less than 2 week old samples of equal size that were differently prepared, show that there is no measurable difference between them (figure 4.1a). Particles that are synthesized, prepared and stored under nitrogen do not differ from equally sized particles that were only synthesized under nitrogen atmosphere and then prepared in contact with air. This seems to be true at least for particles that are not exposed to air for longer than 2 weeks. The comparison of the line shapes of the cobalt L_3 edges, a measure of the oxidation state of the nanoparticle is very similar for the investigated pairs of equally sized but differently stored particles. It therefor seems that the native oxidation can not be reduced by the nitrogen atmosphere during the post synthesis processes and therefore must occur during synthesis.

By comparing on the other hand different measurements of samples stored in air, with unequal sizes or ages, there are strong differences from sample to sample visible (figure 4.1b). While some spectra appear almost without structure in the L_3-edge as to be expected for pure (not oxidized) cobalt, others appear to be nearly fully oxidized showing much fine structure of CoO.

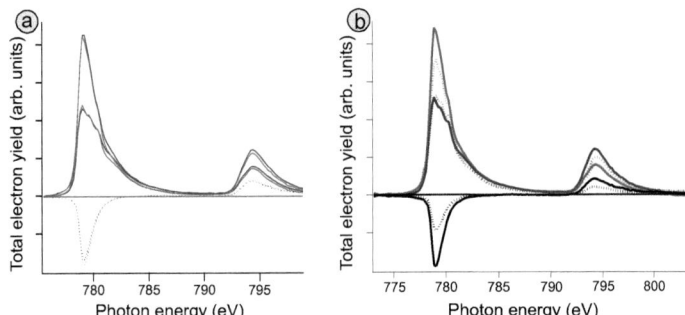

Figure 4.1: Influence of storage conditions in N_2 or air: Two 12 day old similar sized (3.7 nm and 4.0 nm) but differently prepared samples show, that there is no apparent effect of the oxidation state of the sample, whether the entire preparation process is carried out in a nitrogen atmosphere or not. (b) Spectra of two differently oxidized samples (15% solid lines - 50% dotted lines) shows strong differences. There is more fine structure visible in the L_3-absorption edge of the stronger oxidized sample and it's dichroic signal is roughly only half of the other.

These strong differences in oxidation need to be understood. First step is to define a measure to display the oxidation state of each sample and if necessary to investigate what kind of oxide is present. When comparing the fine structure of the spectra of the strongest oxidized samples with spectra from the literature of pure CoO and Co_3O_4 [14, 28] (section 2.2) one can see easily that there is only CoO present in the samples under study. With cobalt being ferromagnetic and CoO anti-ferromagnetic, it makes a big difference in the magnetic properties of the particle. A measure of the oxide in the spectrum can be defined by taking a pure cobalt and a pure cobalt oxide spectrum and subtract their background exactly as all the other samples. After removing the step function and normalizing the areas under the curve to the amount of d-holes of the specimen, in this case using the numbers of d-holes from the literature $n_{d-holes(Co)} = 2.25$ and $n_{d-holes(CoO)} = 2.73$ [103]. Adding these two spectra one creates a simple simulated transition from (oxide free) metallic cobalt to completely oxidized cobalt. The changes at the L_3-edge are much stronger than at the L_2-edge, therefore it is best to concentrate on the features of the L_3-edge. Taking the ratio of the peak height "C" and "B" of two structures of the cobalt oxide spectrum at the L_3-edge from which the position of peak "B" almost coincides with the position of the pure cobalt peak (figure 4.2), one can calculate a unit free and normalization independent measure for the amount of oxide in the sample (figure 4.3). By comparing the ratio of the peaks of the measured samples with my above described scale, one can get an estimation, of how much of the measured signal was coming from oxidized cobalt atoms in the samples.

Displaying the oxidation versus the age of the sample(figure 4.4),

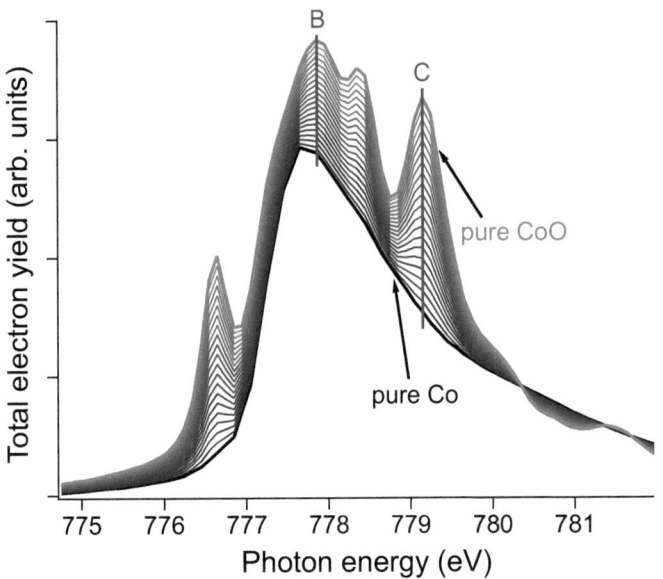

Figure 4.2: By comparing the peak heights of the cobalt oxide structure in the XAS spectrum with the heights of the XAS spectrum of pure cobalt at the same energy position, one can define a rough measure for the amount of oxide measured in the spectrum

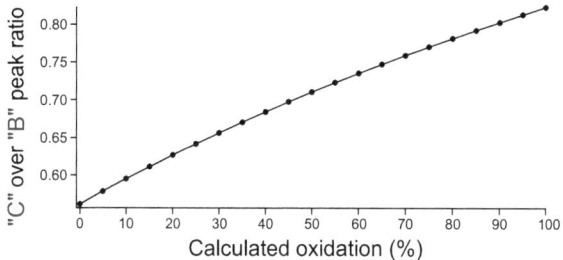

Figure 4.3: Taking the ratio $\frac{C}{B}$ for all combinations from the pure cobalt spectrum to the pure oxide spectrum varying the mixture by 5% from spectrum to spectrum one receives a fairly linear relation between the ratio of the peaks and the amount of oxide in the spectrum

one can easily see a trend that older samples are more oxidized. The oxidation range from 10% to 85% with increasing age, displaying an obvious aging of the samples. The one sample we measured at two successive beam times had been strongly oxidized to start with and shows no further oxidation. For pure 9.5 nm Co nanoparticles it was reported by Wiedwald et al. [149] that oxidation stops after a shell thickness of 2.0 nm to 2.5 nm is oxidized. The oxidation of the here investigated Co_xPt_{100-x} nanoparticles appears to saturate and reach it's maximum after 160-200 days.

Reordering the graph in a way that it displays the oxidation versus the size of the particles it appears that the oxidation is increasing not only with age, but with particle size too (figure 4.4).

Restricting the set of data points to relatively fresh samples of 11 to 15 days of age (figure 4.6a) it appears different. There is no ordering of oxidation that follows the particles age. Displaying the oxidation versus particle size there is an trend noticeable (figure 4.6b). The

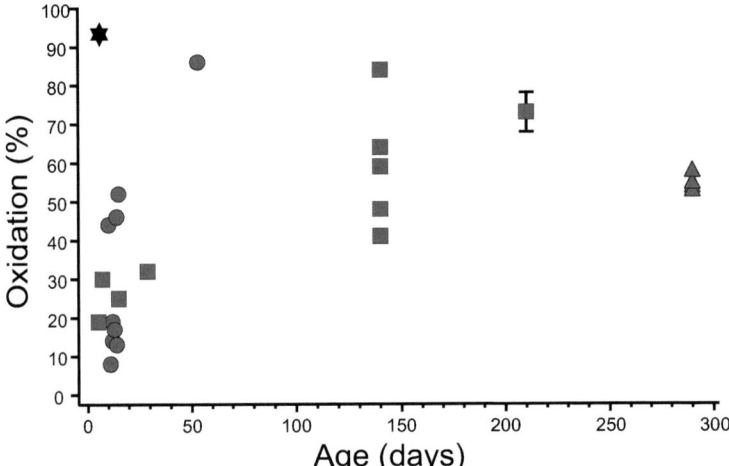

Figure 4.4: Displaying the oxidation of the different measured samples against the age of the samples at the time of the measurement, shows a distinct increase of the average oxidation with time, which is combined with a broad spreading of the oxidation of similar aged samples. The brown circles, squares and triangles represent different beamtimes, while the black star is a synthesis that was already known to have been gone fail, due to a different color of the synthesis solution.

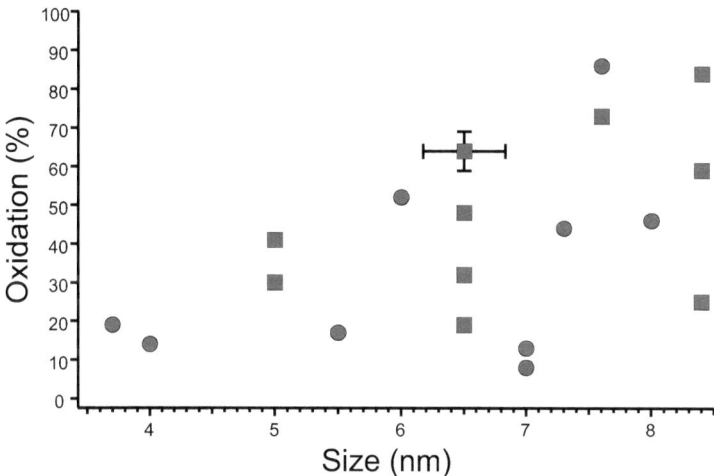

Figure 4.5: Displaying the oxidation of the different measured samples against the size of the samples, it appears that bigger nanoparticles tend to be more oxidized.

larger the particles the weaker they are oxidized. This is the inverse trend compared with that of all particles measured and including all differently old samples measured. If one assumes that all surface cobalt atoms are oxidized, which is a good estimation, since the organic ligands are attached to the outer shell cobalt atoms. It is evident that the most outer shell is entirely consisting of oxidized or otherwise covalently bound cobalt atoms, leading to a Co^{2+}-state. Remembering that during the wet chemical size selective precipitation not all precursors were removed, there should be at least some precursors (containing Co and probably being oxidized) on the surface of the particle. All particles are to first order spheres and if one compares the surface volume of one atom thickness to the remaining volume of the particle, one will see that it follows the same trend: The bigger the atom the smaller the ratio of surface to volume. This way one can understand that in the case of "fresh" samples the lager the size of the particle, the smaller is the amount of oxidized surface atoms in comparison to all atoms present, explaining the trend in the graph 4.6(b).

At least parts of the particles age and are not stable in time. The oxidation is largest for the smallest fresh samples and inverts with time so that the largest samples are more strongly oxidized. This effect shall now qualitatively and quantitatively explained in a model (figure 4.7). The organic ligands (mainly carbon chains) are assumed to be transparent for the soft x-ray light. The light has a penetration depth of 250 nm in cobalt bulk at the cobalt L-edges. The exponential attenuation due to absorption of the incoming light by the already penetrated particle leads to an intensity loss of the exiting soft x-ray light on the backside of a 9 nm particle is roughly

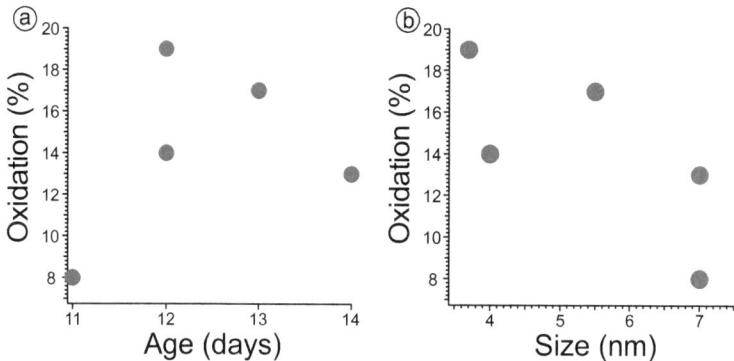

Figure 4.6: Displaying Oxidation of similarly fresh samples versus age (a) and versus size (b), it is quite striking that the more obvious change seems to be correlated to the size, but not the age.

5%. Additionally some assumptions were made: The particles are considered perfect spheres, while the cobalt atoms appear evenly distributed over the entire sphere as a Co-density cloud. The same absorption cross section is valid for all cobalt atoms (approximating $\sigma_{CoO} = \sigma_{Co}$). All stimulated cobalt atoms emit the same amount of electrons with a spherical distribution, thus assuming randomly orientated atoms in the particle and circularly polarized light. Adding up all electrons that will be able to escape the particle from the point of there excitation, one gets a measure that is similar to the total electron yield used in the experiments. Using the sampling depth for the electrons of 25 nm for cobalt bulk material [99], roughly 15% of the most inner electrons exited in a 9 nm particle will not leave the particle. Similar as for the light absorption, the real sampling depth differs slightly for cobalt and cobalt oxide, but is neglected in this model. The experimental data provides an information about the relative amount of cobalt oxide to metallic cobalt, therefore the values calculated by the described model should be the ratio of electrons escaping the sample originating from all oxidized cobalt atoms to the amount of electron that escape originating from all metallic cobalt atom. The chemists found a trend in there synthesis, that smaller particles (3 nm) consist of $Co_{25}Pt_{75}$ and bigger ones (9 nm) of $Co_{50}Pt_{50}$. This was included into the model by splitting the particle in a core of $Co_{25}Pt_{75}$ up to 3 nm in diameter and a hull of increasing Co content, which is highest at the surface. Since it is most unlikely that the particles completely restructure during the synthesis all the time to keep a constant stoichiometry in every volume element of the particle, the following procedure was chosen for the model: The change from $Co_{25}Pt_{75}$ at 3 nm to $Co_{50}Pt_{50}$ at 9 nm was assumed linear, while a radius dependent Co_xPt_{x-100} stoichiometry was calculated. Since the Co content increases with particle size, the outer shells contain even more Co than the particle average.

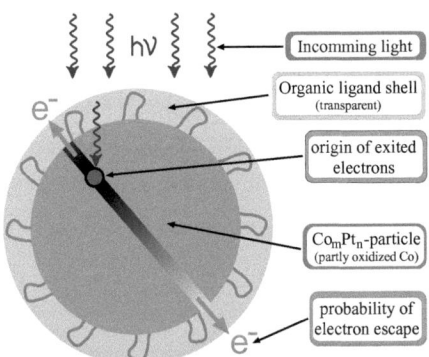

Figure 4.7: The simplest model would be taking the particle as completely homogeneous and letting it oxidize homogeneously

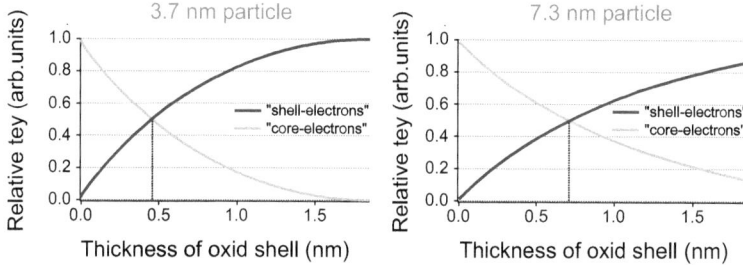

Figure 4.8: Model calculation show the portion of the tey measured due to shell electrons (from the cobalt oxide) and the portion of the tey measured due to core electrons (not oxidized Co), all in dependence of the oxide shell thickness. This is done for two particle sizes (3.7 nm and 7.3 nm). Almost the doubled thickness of the oxide shell is required to attribute raise the shell portion of the tey to 50%, which mostly is a consequence of the decreasing surface to volume ratio of larger particles.

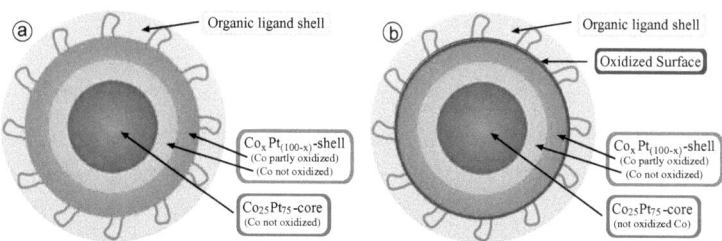

Figure 4.9: The simple idea of a homogeneous particle is extended in two steps (a) with a $Co_{25}Pt_{75}$ core and maximum oxidation depth and (b) a completely oxidized surface

Table 4.1: Measurable oxidation in fresh nanoparticles is compared with model calculations. A good agreement can be found, when a solid shell of 0.075 nm of the particle is oxidized already in the synthesis process. Since the diameter of a Co atom is about 0.2 nm the smaller model value may be reached, when taking into account that organic ligands are attached to the outer shell and that this outer shell may therefore not be solidly filled.

size of particle	3.7 nm	5.5 nm	7.0 nm
particle oxidation (measurement)	19%	17%	11%
surface oxidation 0.075 nm (model)	16%	13%	11%
surface oxidation 0.100 nm (model)	21%	16%	15%
surface oxidation 0.150 nm (model)	30%	24%	21%

The oxidation of a particle will start at the surface and head inwards. The topmost layer is already affected during synthesis and may not be completely filled with Co and Pt due to the attached organic ligands. The organic ligands covalently bind to Co surface atoms using $Co-OOCR$ bonds, leaving the involved Co atom in a Co^{2+} configuration as in cobalt monoxide. This surface oxide had qualitatively already been shown for earlier synthesized batches of these particles with PES spectroscopy by the group of T. Möller [88]. While testing the derived model several calculations were made leading to the best results, when considering the topmost layer completely oxidized, but only two fifth filled with metal. This is attributed by considering the outermost layer as 0.075 nm thick (diameter of a Co atom≈0.2 nm). As can be seen in table 4.1 the measured oxidation state of fresh nanoparticles was between 10% and 20 %, where the smaller particles are more strongly oxidized.

If one assumes all particles to oxidize completely, one would end up with 100% oxidized particles, of which the small particles reach this state of complete oxidation first. Obviously this scenario does not apply here[1] since it does not explain either of the two observations: The oxidation appears to saturate and in the final state the large particles are more oxidized than the small. The next idea would be to assume the particles to oxidize only a certain depth from the surface, as reported for the pure Co cluster by Wiedwald et al. (oxide shell thickness ≈2.0 nm-2.5 nm) [150]. Applying this idea (shown in figure 4.8) of fully oxidized shells with similar oxidation depth for all cluster sizes in the model, one can see in table 4.2 that again the small particles would have to be oxidized more strongly than the large.

The particles measured in this work consist of CoPt alloy and differently to pure Co nanoparticles it is not to be expected that all cobalt atoms in the CoPt nanoparticles can be oxidized as they would in pure cobalt particles. Cobalt in a cobalt-platinum-alloy is known to be very stable and can not be oxidized under normal conditions (section 2.2.2). Lets assume that everywhere, where it is stoichiometrically possible, the atoms forms small, stable $CoPt_3$ groups that hinder the cobalt in this group to be oxidized, hence allowing only cobalt that is not in a stable $Co_{25}Pt_{75}$ configuration in the particle to oxidize (figure 4.9b). If one calculates the amount of oxidation measured with the model with the outer shell of 0.1nm thickness is completely oxidized and stops after a shell of 1.0 nm

---

1. only one sample was measured to be almost completely oxidized, but this was sample number 125B and its solution had already turned black during synthesis, which should normally not happen and can now be considered as a sign for complete oxidation of the cobalt.

2. As described in the text this shell is only stoichiometrically oxidized, with all Co in a $CoPt_3$ environment remaining metallic, while in the surface layer all Co is oxidized, but only 2/5 of the surface sites are filled with metal atoms.

Table 4.2: Measurable oxidation in aged nanoparticles is compared with different model calculation values. In the model all particles have a 3 nm core of $Co_{25}Pt_{75}$ stoichiometric distribution, hence this core is considered not oxidizable. The absolute errors for the oxidation state in the table is ±5%, while the relative error is below ±2%.

size of particle	5.0 nm	6.5 nm	8.4 nm
measurement	41%	48%	59%
0.075 nm fully oxidized shell	13%	12%	11%
0.250 nm fully oxidized shell	34%	30%	27%
0.500 nm fully oxidized shell	56%	50%	46%
1.000 nm fully oxidized shell	83%	76%	70%
0.5nm shell^2 with fully oxidized surface	34%	37%	40%
0.9nm shell^2 with fully oxidized surface	40%	48%	54%
1.0nm shell^2 with fully oxidized surface	41%	50%	56%
1.1nm shell^2 with fully oxidized surface	—	51%	59%
1.5nm shell^2 with fully oxidized surface	—	54%	64%

±0.1 nm is oxidized, while only the Co not in a stable $CoPt_3$ group oxidizes, the results are much more satisfying. Consider an error of ±5% for the measured oxidation.

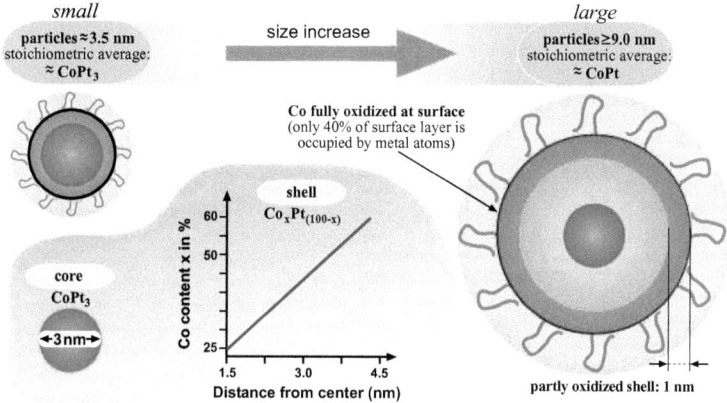

Figure 4.10: Only the seeding process in the synthesis is temperature dependent, therefore all particles will grow during synthesis unaffected by the temperature, hence one may expect that the most inner part of all particles looks alike. The core is in a stoichiometric composition of $Co_{25}Pt_{75}$, while the shell is increasing in the Co content, while approaching the surface. This way the average Co content shifts from $Co_{25}Pt_{75}$ to $Co_{50}Pt_{50}$ from small to large particles, while leaving the cores of all particles unaffected. All Co atoms at the surface that are attached to organic ligands are oxidized during the synthesis, while a shell of up to 1nm may stoichiometrically oxidize over time. The Oxygen can only affect Co atoms that are not in a stable $CoPt_3$ configuration, which leads to a stronger oxidization of larger particles over time.

It therefore appears that the particles are fully oxidized at the surface and will in time oxidize until all cobalt not alloyed as $CoPt_3$ in the outer 1 nm shell is oxidized.

Some pairs of samples consisted out of equally sized nanoparticles, one prepared as single flat layers, the other crystalized to macro-crystals of these particles and then spread over the silicon wafer. All those samples were 10 months old when measured an age at which the flat layered samples have reached their saturation oxidization.

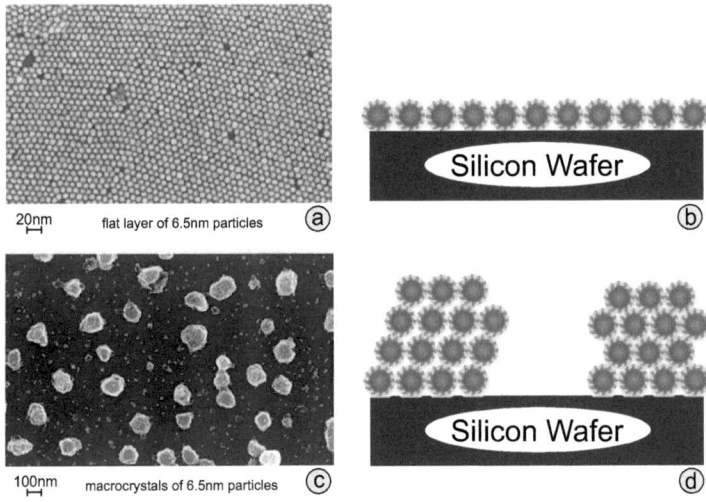

Figure 4.11: Most of the samples were prepared as single layered films (a+b), but some solutions had been treated to form macro crystals and those were deposited as scattered single particles. All preparations were SEM characterized prior to and after the X-ray measurements to investigate any possible beam damage. No beam damage could be found. The SEM pictures were taken by A. Kornowski of the Group of H. Weller of the physical chemistry department of the University of Hamburg.

The 5nm particles show no difference in their oxidation state, whether they are prepared as a flat layer or as a macro crystal. The larger the particles, the bigger the difference of oxidation states of macro crystals and flat layers, up to 25% enhancement for the 8.4 nm macro-crystals (figure 4.12). There is little known about the actual structure of those macro crystals. Assuming the nanoparticles rather round, any kind of packing to construct nanoparticles will leave holes in the crystal that occupy at least 26% of the crystal volume (assuming hcp or fcc). If those holes are filled with oxidized cobalt precursor or other smaller fragments created during the crystallization process, which takes place while the nanoparticles are still in a liquid solvent, this could favor this trend in two possible ways. One would have to assume that whatever fills these holes contains oxidized cobalt and needs a minimum hole size, which is not present in a 5 nm (nanoparticle size) macro crystal. In a model just using spheres and in case of hcp or fcc structure with 5nm diameter Particles the macro crystal would allow spheres of slightly below 0.8 nm diameter to be embedded, with a bcc-structure these spheres could have almost 2.1 nm diameter. This could either lead to more oxidized material measured, if the filling material packs more closely if there is more space available or if the structure of the macro crystals may change to a more loose packing, when increasing the macro-crystal size.

4.2 Branching ratio

The apparent Spin Orbit splitting of the nanoparticles shows no dependencies (figure 4.13a). This is a plausible, since SO splitting is an atomic property and changes little from for example Co to CoO. Quite different is the behavior of the Branching ratio that depends strongly on the core hole screening of the final state after photo absorption (figure 4.13b). For slightly oxidized wetchemically prepared nanoparticles the BR of 2.6 is between that for $CoPt_3$

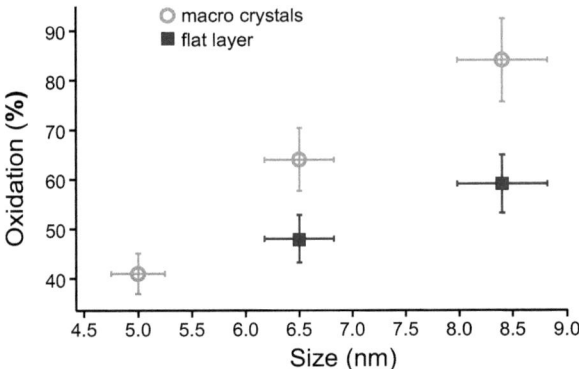

Figure 4.12: Comparing the oxidation state of the standard flat single layer prepared 160 day old samples with macro crystals of the same age, both measurements show a size dependent increase of the oxidation measured, but the macro crystals have a steeper slope and reach up to 35% higher oxidation states.

thin films (2.7) and Co bulk (2.3) and rises up to 2.9 (at 60% oxidation) for the monolayer prepared samples of nanoparticles and further to almost the value of CoO thin films and CoO 2 nm particles (3.7) for the highly oxidized macro crystalline samples and the obviously malfunctioned synthesis that produced 95% oxidized particles (3.6). There is some small step at 60% oxidation in the curve and a slightly higher slope than at lower oxidation states. The step could be caused by the different samples and the different stoichiometric composition that might have been measured, since the nanoparticles prepared as thin films are very similar and contain no spaces in the lattice structure, as might be present for the macro crystals; the 95% oxidized sample (125B) was already addressed to in the oxidation discussion and was strangely black colored in the synthesis process. The slope at higher oxidation follows the linear transition from the BR of unoxidized Co (2.3) to CoO (3.7). It is to be expected that if all Co that can be measured is oxidized that the BR is that of bulk CoO.

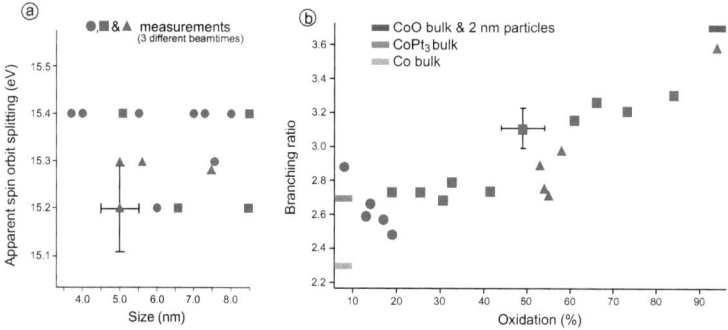

Figure 4.13: The apparent SO-splitting of the particles (a) did not show any dependencies, neither to the oxidation state, nor to the here displayed size. The BR of cobalt bulk (orange bar) [9], of CoPt$_3$ (light blue bar) [53], 2 nm CoO particles and thin films (brown bar) [69, 110] vary strongly. The particles measured in this work fit well into these values and show a clear increasing dependency with their oxidation state, approaching the literature value of 3.7 for 100% oxide.

4.3 Ratio of orbital to spin magnetic moment

Before discussing the magnetic properties, that were extracted using the XMCD sum rule introduced in section 2.1.1 some coments on the T_z term that adds to spin moment in the sum rules: The T_z term (section 2.1.1) is considered to be negligible for Co in all cited reference measurements (table 4.7). As measurements have shown the sum rules may be used with this assumption for Fe, Co and Ni [25, 102, 152, 153], at least as long as the systems provide a rather homogeneous environment for the Co atoms (or provide samples with randomly orientated particles). The measured nanoparticles are large enough that less than 20% of the atoms (for the smallest sizes) are at the surface. The surface is evenly covered with organic ligands and thus it is not to be expected that there is a ligand induced strong deformation of the crystal lattice. Only the deformation upon deposition that was suspected due to the angular dependence measured may lead to some inhomogeneous lattice deformation, but that is not to be expected to be higher than that of thin film systems of late $3d$ transition metals for which the assumption of negligible T_z has been claimed to be shown experimentally. In this case one may not simply assume T_z to be zero as usually done for randomly arranged samples or angle dependent XMCD measurements [34], since the strong spin orbit coupling of the Pt may lead to a non zero contribution of the magnetic dipole term that is independent of the crystal field. Ab initio LSDA calculation [35, 75] show diminishing T_z for monoatomic Co-chains on Pt surfaces. Often theoretical calculations achieve rather bad agreement with the actual orbital magnetic moments since they usually neglect any electron correlation effects, hence there might be still some reasonable doubt, whether or not any claimed diminishing of T_z in a strong spin orbit coupled system will really come to pass. However calculations of Komelj et al. included some electron correlation effects by a so called orbital polarization term [113] and thus even slightly overestimated the orbital magnetic moment of Co chains on Pt(997). Hence for the wet chemical Co_xPt_{100-x} discussed in this chapter T_z will be assumed to be negligible.

For diminishing T_z the XMCD sum rules have a minimum uncertainty, when applying them to the ratio of orbital and spin magnetic moment, because the number of d-holes disappear in that ratio and this number is not precisely known for these particles. This ratio increases with the level of oxidation, which is quite expected, since the orbital to spin moment ratio is three times as high for cobalt oxide as it is for metallic cobalt. There is no reason, why this trend should be much different just changing the size of the particle from bulk to nanoparticles.

A display for differently oxidized samples visualizes the increasing ratio of orbital to spin magnetic moment with increasing level of oxidation (figure 4.14).

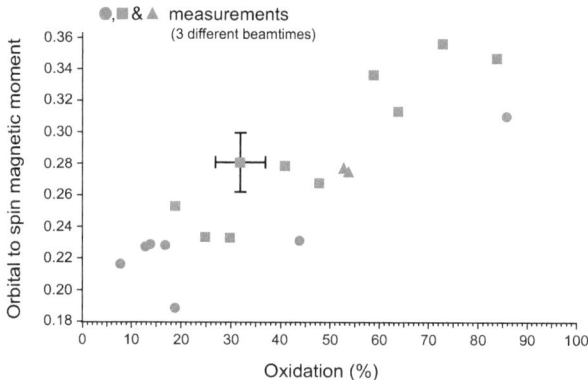

Figure 4.14: The ratio ml/ms apparently depends linearly on the oxidation

Comparing flat layer preparation with the corresponding macro crystals, no real enhancement of the orbital to spin magnetic moment is noticeable in normal measuring geometry, despite of the expectation stirred by the above described increased oxidation of the macro crystal. Above that their is no increase of the orbital to spin moment ratio for a macro crystal if one changes the measuring geometry from normal to the surface to 60° out of normal incidence, while keeping light and magnetic field parallel to each other. The measuring geometry of the used chamber did not allow measurements that involved angles smaller than 30° out of the plane.

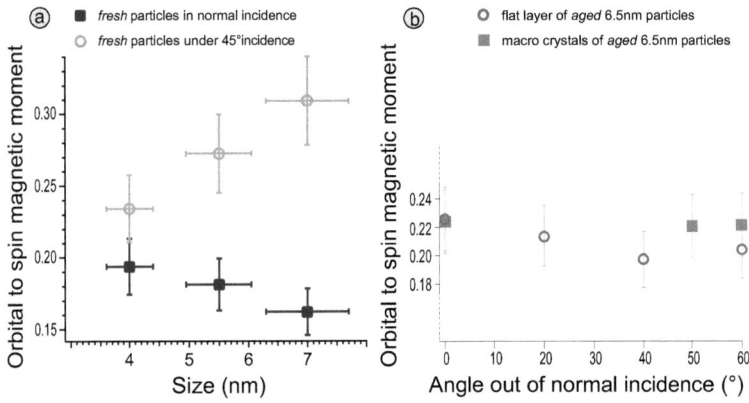

Figure 4.15: Comparing fresh and aged samples, measuring under different geometries shows, that apparently fresh samples show higher ratio of orbital to spin moment under 45° than at normal incidence and an increase of this effect with cluster size. This effect disappears completely if the samples age and there is no difference measurable between flat layers and macro crystals of nanoparticles.

There were quite a few differences and changes noticeable between the different particles and preparations with effects up to 25% of the measured values. A comparison of all measured samples, only distinguishing between lighter (less than 30%) and stronger (more than 50%) oxidized samples, with results of others for cobalt and cobalt-platinum systems measured and calculated is done and illustrated in figure 4.16. The nanoparticles measured in this work show a strong enhanced ratio of orbital to spin magnetic moment compared with

CoPt bulk (about 25%). The enhancement is smaller for the less oxidized particles and stronger for the more oxidized systems. The large deviation between most experimental and theoretical values demonstrates the need for some additions to the theoretical models describing this sort of system.

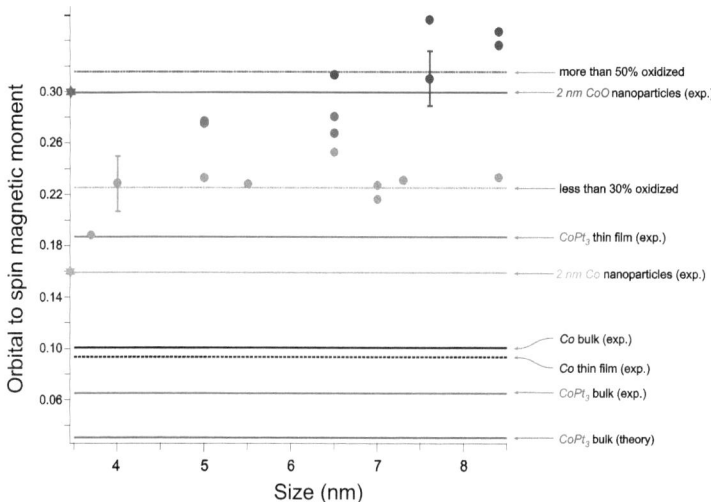

Figure 4.16: The graph shows the results of m_l/m_s and in comparison measurements of 2nm Co and CoO nanoparticles (orange and brown star+line [70]), CoPt$_3$ thin film [53], bulk material of CoPt$_3$ [68]and cobalt [25] and finally with theory [19]. The magnetic properties of the CoPt nanoparticles are strongly enhanced to bulk values, but only 10-15% above thin CoPt$_3$ film and 20% above pure Co 2nm nanoparticles. The same is true for the more oxidized samples in comparison with the pure CoO nanoparticles.

4.4 Orbital and spin magnetic moment

Taking full advantage of the sum rules evaluating spin and orbital moment separately for the different measurements, some results become clearer. We had seen an increase of the ratio ml/ms with oxidation. Taking a closer look at those increases, plotting versus age one sees, that spin and orbital magnetic moments decrease and that there is a wide spreading in those values (figure 4.17).

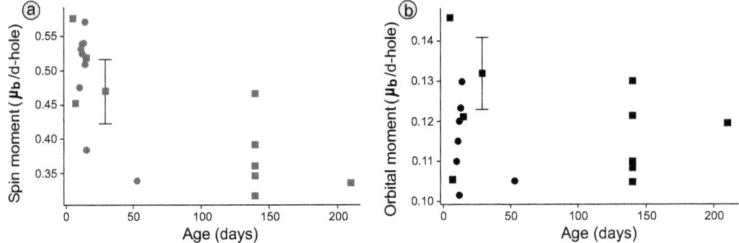

Figure 4.17: Both spin and orbital magnetic moment decrease with age of the particles, since the relative decrease of the orbital moment is less than that of the spin moment, the ratio of the two values ml/ms increases.

This is due to the fact that there were differently strong oxidized samples of the same age measured. Obviously the different samples do not age in the same speed. Displaying spin and orbital moment in respect to the oxidation level, the decrease of the spin moment is very apparent (figure 4.18), while the orbital moments decrease only very little. Please note at this point that all the graphs plot magnetic moments per number of d-holes. The effect on the total measured spin moment is therefore a little less pronounced. The orbital magnetic moment is as well decreasing with oxidation, at least one may believe it with much some good will. This can be explained if one takes a look at the the angle dependent measurements we had performed for fresh particles (figure 4.19).

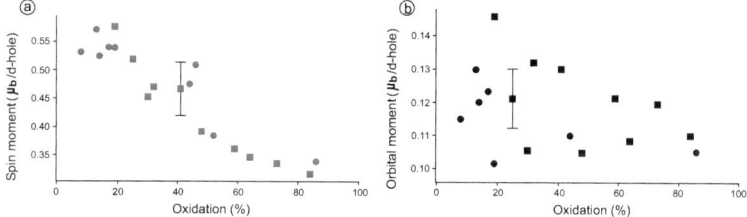

Figure 4.18: Plotting spin and orbital magnetic moment versus oxidation, one can see very nicely that the spin moment reduces linearly with the oxidation, while the orbital moment is decreasing too, but shows a stronger variation for less oxidized particles. This broadening in values for little oxidized particles is due to the fact that these particles have a size dependent orientation on the surface, where larger particles have their magnetic ordering more in plane than perpendicular to the surface.

It is nicely to be seen that there are no changes in the spin moment, which is to be expected, since the external magnetic field of 6 T is enough to saturate the spin moment in any direction of the sample. The spin moment is not directly coupled to the crystal field. The orientation of the spins is without external field only due to spin orbit coupling. The spins remain rotatable and can be aligned to an external magnetic field. The orbital magnetic moments are less flexible in their orientation, since the orbital's of the atoms are aligned to the crystal structure of the nanoparticle. When increasing the cluster size there is a decrease of the orbital magnetic moment in the normal to the surface geometry and an increase when measuring in 45°angle to the surface normal (figure 3.17). This implies that there must be some sort of orientation of the clusters on the surface that favors in plane magnetization with growing cluster size. This could be achieved, if the particles deform becoming flat (figure 3.6) when coming in contact with the surface, instead of having a spherical shape. This would lead to some shape anisotropy and one would expect an increase of the magnetic moments parallel to the surface plane.

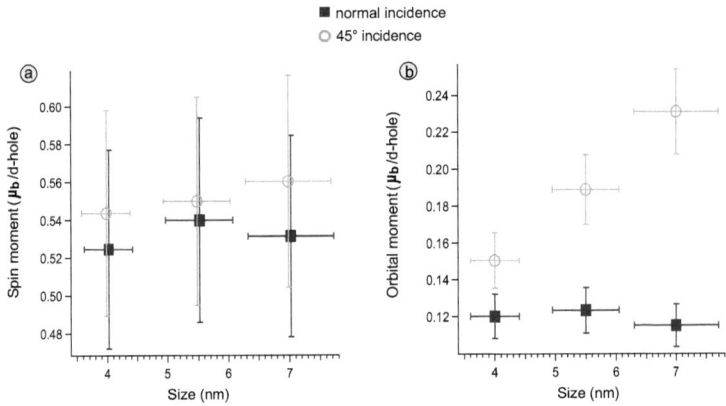

Figure 4.19: It is quite obvious that the measured angular dependence of the ratio between orbital and spin moment are only due to a change in the orbital moment

The measurements plotted in figure 4.18b are all in perpendicular measuring geometry and show a broad spreading in the orbital magnetic moment for weakly oxidized particles, which is due to the size dependence of the orbital magnetic moment shown before. This size dependence disappears with growing age of the samples and thus with the oxidation state of the sample, as angle dependent measurements of older particles have shown (figure 4.20). The round symbols in the figure represent a preparation as a flat film, the squares a sample preparation as macro crystals. There is a decrease of spin and orbital magnetic moment to be seen for the thin film preparation, but rather a constant behavior for the macro crystals. This is both different to the fresh 5nm nanoparticles, where an increase of at least 20% at an 45°angle was measured (figure 4.19). The older particles have several layers of oxide on the surface of the particles and cobalt oxide is an anti-ferro-magnet. In XMCD measurements only the net magnetic moments will be visual, thus changing cobalt to cobalt oxide will reduce the net measured magnetic moment of the sample. The slight decrease of spin and orbital magnetic moment for the flat layer preparation is due to the increased amount surface region (containing cobalt oxide) that will be measured, when decreasing the incidence angle of the light. The illuminated surface area of the sample will be increased, while the penetration depth perpendicular to the surface decreases. This argument will not apply, when having blocks (as macro crystals) on the surface that are well separated, since in that case the light will just illuminate the sides of these blocks, when the angle is tilted, but the penetration depth perpendicular to a surface stays almost constant.

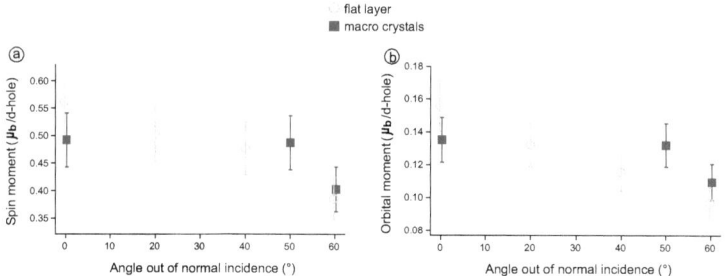

Figure 4.20: When inspecting spin and orbital magnetic moment of older 5nm nanoparticles prepared as flat layer and as macro crystals, there is no angular effect measurable.

4.5 Discussion

Experiments on free iron, cobalt and nickel clusters in 1994 [16] have shown that the typical size regime in which clusters resume the total magnetic moment of the corresponding bulk material is between 300 and 500 atoms per cluster and thus in the order of a 2 nm diameter particle. When nanoparticle are deposited on a surface, there are differences in the photo absorption measurable for 10nm CoO particles, that disappear at 20 nm particles [129] and were explained as surface defect related. Deposited 11.4 nm partially oxidized wet-chemically prepared nanoparticles have shown strong enhanced m_l/m_s ratio [150] and mass filtered 5 nm to 12 nm deposited Co nanoparticles have displayed highly reactive behavior in contact with oxygen [12], which will not occur at Co-film surfaces [7]. The particles chemically prepared were all in the size regime between 3.7 nm and 8.4 nm, hence the expectations were to find some, but not very strong magnetic size effects. The measured CoPt nanoparticles show magnetic properties that are very similar to 2 nm $CoPt_3$ nanoparticles measured by P.Imperia et al. [70], ranging from ml/ms=0.16 for not oxidized particles to ml/ms=0.30 for completely oxidized particles. A measurements performed by P.Imperia at 2 nm sized CoPt nanoparticles with different stoichiometric ratios lead to the conclusion that the orbital to spin ratio of the magnetic moment in CoPt nanoparticles shift to higher values with the increase of cobalt in the nanoparticles. The same shift is to be seen in the here performed measurements when comparing similarly slightly oxidized samples of different size and thereby different stoichiometric ratio of cobalt and platinum (figure 4.19). The Particles show a angular dependence in their ratio ml/ms, but with a stronger increase for in plane orientation, as the decrease of the out of plane orientation. Hence there is a net increase of the ratio with particle size and thus higher amount of cobalt in the sample. On the one hand this is nice, since it shows a consistence in results between two very differently prepared sets of samples, on the other hand the possibility that there might be any size dependence magnetic properties is reduced to the angle dependence of the fresh nanoparticles. Angle dependence was not expected. When measuring randomly distributed particles on a not or just slightly interacting surface (as organic ligand covered particles on a silicon surface) one expects no preferred orientation of the particles and hence no changes, when changing the measuring geometry. The size dependent changes that appear when measuring fresh nanoparticles under different geometries show that there must be some interaction between particles and surface, most likely the deformation of the particles on the surface, followed by magnetic shape anisotropy. There could be a ferromagnetic particle-particle interaction that leads to this ordering, which disappears when the particle shell oxidizes and becomes thereby non,- or anti-ferromagnetic. The bigger particles are a little bit less spherical, the largest sizes appear as squares in the TEM and SEM pictures, this non spherical shape can lead to form anisotropy and can favor a certain in plain magnetic magnetization, which disappears when the samples oxidizes, due to the anti-ferromagnetic state of CoO. The oxidation state that is reached at Co particles of 5 nm to 12 nm diameter, that were prepared oxide free under UHV-conditions and later exposed to 500 L of oxygen[3], as performed in the Group of K.H. Meiwes-Broer [12], will be reached by the CoPt nanoparticles measured in this work after about 200 days of storage in contact with air. This shows that the oxidation process is slowed down drastically by using a CoPt alloy instead of pure Co.

Table 4.3 shows all magnetic moments derived from the measured data of all particles that were freshly prepared and just slightly oxidized. As described in section 2.1 about XAS, the tey suffers self absorption effects when the material exceeds the thickness of the escape depth of the electrons and if the absorption length of the x-rays is significantly larger. This is the case, since the escape depth of electrons in cobalt is slightly above 2 nm and the penetration depth of the x-rays is roughly 10 times as much. Extracting the correction factors from K.Fauth [41] for the sizes used in this work, table 4.4 was derived. When applying these corrections, the spin

---

3. 1 L=10^{-6} Torr sec, hence is a dose of 500 L oxygen equivalent to the exposure to air for only 10 μsec

size/nm	$m_s/(\mu_b \cdot d_h)$	$m_l/(\mu_b \cdot d_h)$	m_l/m_s	$mag_{tot}/(\mu_b \cdot d_h)$	oxidation/%
3.7	0.53	0.09	0.16	0.62	19
4.0	0.51	0.10	0.19	0.61	14
5.5	0.52	0.09	0.18	0.61	17
6.5	0.55	0.11	0.20	0.66	19
7.0	0.51	0.08	0.16	0.59	7
7.0	0.55	0.09	0.17	0.64	13
8.4	0.50	0.08	0.17	0.58	25

Table 4.3: Magnetic moments of fresh (only slightly oxidized) Co_xPt_{100-x} nanoparticles as measured.

moments change little, but the orbital moments increase strongly and especially change relative to each other.

The result of the corrections applied is displayed in table 4.5. All values appear to be rather similar with maybe a small deviation for the smallest particle size. Free clusters reach bulk like values at sizes of about 3nm diameter, therefore it is not too surprising that there are no size dependent effect visible in the size range from 4-8nm. This has a good side, if the particles are to be used in any application the properties of a product being constructed with 8.4 nm particles will not differ from one that is being built up using only 4 nm particles. The higher stoichiometric ratio of platinum in smaller nanoparticles reduces the average magnetic moment per unit volume, since the magnetism is mainly attached to the cobalt. Please remind that there were some yet not understood angular size dependent properties measured for the orbital moment of the 4.0 nm, 5.5 nm and 7.0 nm sized particle samples.

A comparison with literature values shows nice coincidences for the

size/nm	m_s-corr	m_l-corr
3.7	0.98	0.84
4.0	0.98	0.83
5.5	0.97	0.77
6.5	0.96	0.74
7.0	0.96	0.72
7.0	0.96	0.72
8.4	0.96	0.68

Table 4.4: self absorption correction factors for spin and orbital moments

Co spin moment, where values of 0.56 to 0.6 μ_B per d-hole of 7.5 nm to 8.0 nm pure Co clusters where presented [12, 13, 73] which is about 10% above the measured values of the wetchemically prepared Co_xPt_{100-x} particles presented here. If considering the oxidation state of roughly 10-20% and cobalt oxide to be an anti-ferro-magnet in bulk material this slight reduction in the spin moment appears reasonable. As can be seen in table 4.6 the measured spin moment per d-hole drops drastically when the oxidation state of the particles increase. The other partially analogous reference samples of 5Å to 7Å sized Co islands on Au [8] are slightly higher with 0.58 and 0.70 μ_B per d-hole, but again in the similar regime. Most other measurements that could be found related to the here presented data are cited in table 4.7, but they are to a large extend bulk or thin film systems or limited to only one reference value. Please note that all data in table 4.7 was converted into values per d-hole, if the authors had given the estimated number of d-holes used in their

size/nm	$m_s/(\mu_b \cdot d_h)$	$m_l/(\mu_b \cdot d_h)$	m_l/m_s	$mag_{tot}/(\mu_b \cdot d_h)$	oxidation/%
3.7	0.54	0.10	0.19	0.64	19
4.0	0.52	0.12	0.23	0.64	14
5.5	0.54	0.12	0.23	0.66	17
6.5	0.58	0.15	0.25	0.72	19
7.0	0.53	0.11	0.22	0.65	7
7.0	0.57	0.13	0.23	0.70	13
8.4	0.52	0.12	0.23	0.64	25

Table 4.5: Magnetic moments of Co_xPt_{100-x} nanoparticles tey self absorption corrected.

publication. In most cases as in the here presented measurement the actual number of d-holes is unknown, nevertheless most authors that were in need of the d-holes chose 2.49, which is the arithmetical average of theoretical Co bulk values from ref. [54] and ref. [152]. In order to be able to compare the actual measurements without the influence of the assumed number of d-holes (ranging from 2.25 to 2.8), this number of d-holes was therefore removed if possible. The orbital moments of equally sized (7.5 nm to 8 nm) Co reference particles are only half as big as the measured Co_xPt_{100-x} samples. It nicely shows the expected enhancement of the orbital moment due to the Pt content of the particles compared to pure Co particles. For comparison there are some pure Co clusters of 4 nm and 8 nm on Si wafers [70], where only the orbital to spin moment is known and with 0.1 about half of that of the measured Co_xPt_{100-x} samples, which fits well to the doubling of the orbital moment, when changing from pure Co to Co_xPt_{100-x}.

The XAS and XMCD measurements have shown an initial oxidation of the particles of 10-20% occurs already during the synthesis process and that short term storage in air (several days) does not significantly increases this oxidation. The initial oxidation state decreases with increasing size of the particle and thus reflects that roughly 50% of the surface layer Co of the particle is oxidized during synthesis. The maximum oxidation state of the particles depends on the particle size as well, since the maximal oxidation depth has shown to be almost constant at 1nm. The oxidation speed and total oxidation is vastly reduced if comparing the particles with equally sized pure Co nanoparticles that oxidize almost instantly with only a dose of 500 L oxygen, ending up with twice as thick oxide shells [150]. The wetchemical Co_xPt_{100-x} particles stored in air or in liquid solvents oxidize in half a year, which unfortunately for any possible application is still too fast, although it proofs slow enough for investigation measurements as done in this work. In order to achieve 10 year time stable magnetic storage media grains much needs to be done to ensure oxidation stability of the particles. This could be achieved by a noble metal coating or any other surface layer, that hinders Oxygen to diffuse into the nanoparticle. The magnetization of the particles depends mainly on the oxidation state of the particle. The total magnetic moment per volume of the particle increases with size, since the stoichiometry of the particles change and the Co content increases with size. The only trend visible for a Co atom is for fresh particles, where an enhanced in plane orbital magnetic moment can be seen for increasing particle size. This size dependence disappears when the particles oxidize, while the other magnetic properties of aged nanoparticles remain similar when comparing different sizes, of equally oxidized particles (table 4.6).

size/nm	$m_s/(\mu_b \cdot d_h)$	$m_l/(\mu_b \cdot d_h)$	m_l/m_s	$mag_{tot}/(\mu_b \cdot d_h)$	oxidation/%
6.5	0.39	0.10	0.27	0.45	48
7.6	0.34	0.12	0.36	0.41	73
8.4	0.36	0.12	0.34	0.43	59

Table 4.6: Magnetic moments oxidized wet chemicals tey self absorption corrected

sample	$n_{d-holes}$	$\frac{m_s}{\mu_B \cdot d_h}$	$\frac{m_l}{\mu_B \cdot d_h}$	$\frac{m_l}{m_s}$	reference
$Co_{(2-10ly)}$/Pt(111)				0.2	[134]
$Co_{(5Å)}$/Au/W(110)	2.6	0.58	0.12	0.21	[8]
$Co_{(7Å)}$/Au/W(110)	2.6	0.70	0.17	0.24	[8]
$Co_{(5Å\&Au-cap)}$/Au/W(110)	2.6	0.77	0.11	0.14	[8]
$Co_{(7Å\&Au-cap)}$/Au/W(110)	2.6	0.58	0.09	0.15	[8]
$Co_{(7.5nm)}$/Ni(111)	2.49	0.60	0.05	0.08	[12]
$Co_{(7.6nm)}$/Ni/W(110)	2.49	0.60	0.05	0.09	[73]
$Co_{(8nm)}$/Au(111)	2.49	0.56	0.08	0.14	[13]
$Co_{(11.4nm)}$				0.24 ±0.06	[150]
Co_{2000} in Cu				0.16	[32]
Co_{3600}/Au				0.17	[30]
Co_{12000}/Au				0.12	[30]
$CoPt_{(3nm)}$ [as prep.]	2.628	0.647	0.046	0.071	[141]
$CoPt_{(3nm)}$ [annealed]	2.628	0.727	0.068	0.094	[141]
$Co_{(4\&8nm)}$/Si_{Wafer}				0.10	[70]
$Co_2Pt_{1(2nm)}$/Si_{Wafer}				0.12	[70]
$Co_1Pt_{2(2nm)}$/Si_{Wafer}				0.14	[70]
$CoPt_{(2nm)}$ [in MgO]	2.49	0.185	0.032	0.18	[12]
$CoPt_{(2nm)}$ [in Nb]	2.49	0.080	0.024	0.28	[12]
$CoPt_{(40nm-film)}$	2.628	0.75	0.099	0.13	[52]
$Co/Pt_{(multilayer)}$	2.49	0.627	0.052	0.08	[24]
$CoPt_{3(film)}$	2.25	0.711	0.058	0.19	[53]
$CoPt_{3(bulk)}$				0.26	[68]
$CoPt_{3(bulk)}$				0.094	[145]

Table 4.7: Literature references of experimental values of magnetic moments of Co and CoPt clusters, layers and bulk systems

sample	$n_{d-holes}$	$\frac{m_s}{\mu_B \cdot d_h}$	$\frac{m_L}{\mu_B \cdot d_h}$	$\frac{m_L}{m_s}$	reference
Co$_{(bulk)}$ [LSDA+DMFT]		$1.614/d_h$	$0.138/d_h$	0.09	[145]
Co$_{(wire)}$/Pt(997) [LMTO+OPT]		$2.09/d_h$	$0.86/d_h$	0.41	[75]
CoPt$_{(bulk)}$ [LSDA+DMFT]				0.139	[145]
CoPt$_{(bulk)}$ [FP-LMTO LSDA]	2.628	0.681	0.041	0.061	[49]
CoPt$_{(bulk)}$ [FP-LMTO GGA]	2.628	0.696	0.033	0.048	[49]
CoPt$_{3(bulk)}$ [LSDA+DMFT]				0.099	[145]
CoPt$_{3(bulk)}$ [FP-LMTO LSDA]	2.651	0.694	0.018	0.03	[49]
CoPt$_{3(bulk)}$ [FP-LMTO GGA]	2.651	0.720	0.022	0.03	[49]
Co$_{(bulk-hcp)}$	2.43			0.03	[54]
Co$_{(bulk-hcp)}$	2.55	0.60		0.03	[152]
Co$_{(bulk-sufacelayer)}$	2.60	0.62		0.03	[152]
Co$_{(bulk-fcc)}$		$m_{tot}=1.64\mu_B$			[95]
Co$_{(bulk-bcc)}$		$m_{tot}=1.73\mu_B$			[95]
Co$_{(bulk-hcp)}$		$m_{tot}=1.63\mu_B$			[95]

Table 4.8: Literature references of theoretical values of magnetic moments of Co and CoPt systems

Chapter 5

Mass Selected Clusters

In this chapter the measured data of the mass selected deposited Co_nPt_m clusters will be presented and discussed. The focus of attention is in finding differences in the cluster properties due to cluster size, stoichiometric composition and upon changing of the substrate. The substrates are a strongly interacting 3 ML thin Fe film on Cu(100) and a much weaker interacting 20-30 ML Ni film on Cu(100). Both thin film systems support perpendicular magnetization of the magnetic film. At this point the experimental setup limits the substrates to magnetic systems, that magnetize the clusters via exchange coupling, in order to perform measurements of magnetic cluster properties. An external magnetic field of 5-10 Tesla that would be required to magnetize the small clusters sufficiently could not be applied in the standard spectroscopy chamber, hence a new chamber is now under construction. In the beginning of the chapter the iron thin film substrate is discussed and the treatment of the iron EXAFS-background in the cluster measurements is presented. Additionally in this context interesting reactions of the iron thin film upon cobalt cluster deposition are presented.

5.1 Iron substrate

A 2.5 to 3.5 layer iron thin film was evaporated at low temperatures on the copper crystal to be the perpendicularly magnetized support of the deposited clusters. To separate the signal of the substrate from the clusters. Typically the background was measured either on a fresh prepared film or with a cluster sample, but far away (2-3 mm) from the cluster spot. The XAS spectra of the clusters had shown strong EXAFS-oscillations of the iron film in the region of the cobalt L-edges (section 3.3.3). To remove the iron EXAFS-oscillations from the measured cluster spectra, some clean iron surfaces were prepared and measured. This procedure had successfully been applied for nickel thin film systems in the past. It proved impossible to match the iron background to the measured cluster spectra. A similar misfit of the iron EXAFS oscillation appeared, when using the background measurements that were taken away from the cluster spot. As can be seen in figure 5.2 the iron EXAFS-oscillations have some energy dependence that differs with the cobalt coverage of the iron thin film. It is known for a 3 layer iron thin film grown on a Cu(100) crystal that the magnetic properties change from out of plane to in plane, when depositing half a layer of cobalt onto the iron thin film [29]. This shows that there is strong interaction between the deposited cobalt and the underlaying iron. EXAFS measurements of iron thin film with increasing cobalt coverage has not been reported in literature so far. Figure 5.3a shows the iron EXAFS vs energy and figure 5.3b the tey vs the $\sqrt{Energy}$ (which is $\approx \lambda^{-1}$). An electron with the kinetic energy of 800 eV has a speed of $17 \cdot 10^6$ m/s which is less than 6% of the speed of light, hence the electron can be treated as non relativistic and the de-Broglie wavelength of the electron follows equation 5.1:

$$\lambda_e = \frac{h}{\sqrt{2 \cdot m_e \cdot E_{kin}}} \quad (5.1)$$

The scattering of the electrons by the surrounding atoms that produce the EXAFS oscillations therefore depend on 1 over square root of the kinetic energy. The periodicity of the oscillations is due to the local environment of the scattering atoms. However from 2 or 3 oscillations, as in figure 5.3b, it is impossible to reconstruct distances reliably. For quantitative analysis more measurements would have to be performed extending the covered energy region 100 eV up to the Cu L-edge, which limits the measuring range.

In detailed investigations of Marangolo et al. [92] of NEXAFS and magnetic NEXAFS of thin iron films on Cu(111) distinct changes could be measured in the iron film coverage up to 2.2 ML (figure 5.1). They claimed to be able to directly link the coverage to the shape of the NEXAFS oscillations.

These changes could be explained by the different growth modes of iron on Cu(111), very important seemed the lattice relaxation process in the growth mode close to the coverage of 0.7 ML of iron. The relaxation lead to a shift in the EXAFS oscillations towards higher energies of roughly 10 eV. In our measurements the shift is towards lower energies, but about the same order of magnitude, when adding 0.3 ML of cobalt onto the iron thin film. This substantial influence that the cobalt inflicts upon the iron thin film must be in a long range order. While the EXAFS signal represents the local surrounding of the surface atoms, the measured tey integrates over the entire illuminated spot, since the effects are visible in the tey, the influence

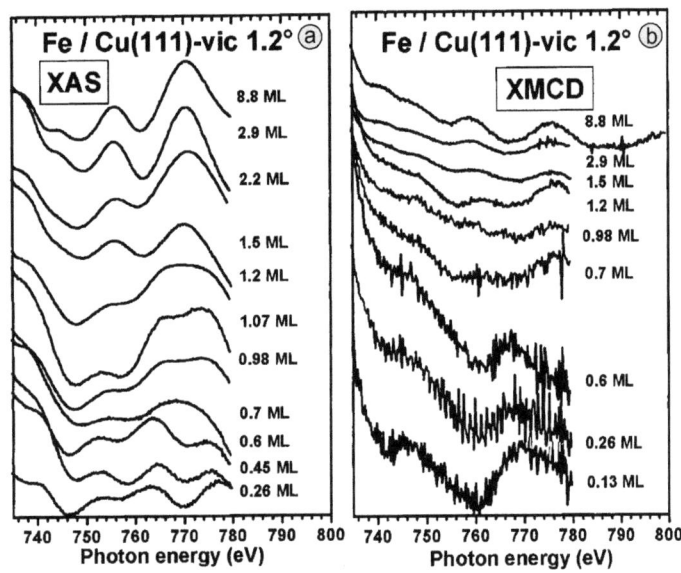

Figure 5.1: NEXAFS and magnetic NEXAFS of iron on Cu(111)(figures from reference [92]) show that there are distinct changes in the spectra, that can be linked directly to the iron film thickness.

of the deposited cobalt must be similar for most iron surface atoms. The shift towards lower energies suggest that the cobalt increases the surface strain, which is consistent with observations of Shen et al. [108, 122, 110]. They observed that adding 0.1 ML of cobalt to an 3 ML iron thin films grown at low temperatures (LT) on Cu(100) started the spin reorientation transition of the film. The effect of the Co capping layer was similar to 10 times that of iron, 0.2 ML of cobalt changed the iron film properties as 2 ML of additional iron would have done. This result suggest that our 3 ML fe film with 0.25 ML Co capping should be similar to a 6 ML fe thin film. Shen et al. also compared room temperature (RT) grown films with films grown at LT and observed that the enhanced surface roughness of the LT grown films lead to a SRT that occurs 1.5 ML earlier than that of the smother films. The magnetism of the iron thin film changed strongly with the adding of the 0.25 ML of cobalt. Using the sum rules (section 2.1.1) the magnetic moments of the iron film with and without the 0.25 ML Co were evaluated and are shown in table 5.1. The clean film has a spin magnetic moment per fe atom, that is in good agreement with XMCD measurements of D.Schmitz et al. [117]. They estimated a spin moment of 2.8 μ_B for a 3 ML iron thin film, that dropped to 0.8 μ_B for 6 ML thick iron films. As mentioned earlier the Co-capped film should behave similar to a 5-6 ML thin fe film, comparing our results with Schmitz et al. the 3 ML films show the same Spin moment and the Co-capped film is with 0.6 μ_B slightly lower, which is to be expected, since the magnetization tilts from out of plane at 2.5-3.5 ML to in plane for higher coverage and the measurements of this work have been detecting the integrated out of plane fraction of the magnetic moments only.

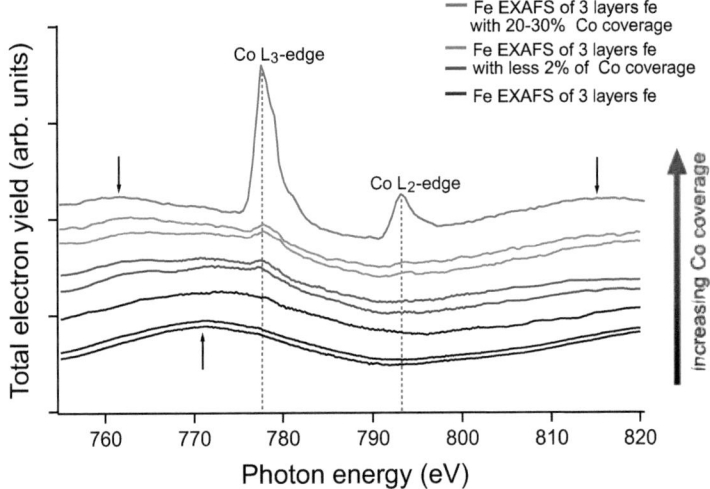

Figure 5.2: The background measurement in the cobalt L_3-, and L_2-edge region show a strong iron EXAFS oscillation. Comparing several measurements of the cobalt $L_{2,3}$-edge region, one can clearly see that there is a strong change in the spectra, which is correlated to the amount of cobalt, that covers the iron layers. Since the EXAFS oscillation seam to shift in their energy position to lover energies with increasing cobalt coverage, the background of the clean iron films can therefore not be used at the coverage of up to 5% of cobalt which is used for the cluster measurements.

The measured background signals displayed that the signal of the oscillation is smooth. There are no additional absorption features to be seen. This was different in the last beamtime that added data to this work and in which the clusters on the magnetic nickel thin film were measured, where the switching mirror unit had been plasma cleaned to remove carbon contaminations prior to the beamtime and where the high frequency antenna to ignite the plasma had a brass socket, leading to different contaminations in the beamline (including a little Co). Since the measured background was smooth with no additional absorption features in the case of the iron thin film substrate, the measured spectra were fitted. The fit was split into the pre L_3-edge, the post L_2-edge and in the region between the two absorption peaks and was smoothly extended by a polynomial

$\frac{film-thickness}{ML}$	m_s μ_B	m_l μ_B	$\frac{m_l}{m_s}$	mag_{tot} μ_B	
3.4 fe (clean)	2.84	0.17	0.06	3.01	this work
3.4 fe + 0.25 Co	0.59	-0.01	-0.025	0.57	this work
3.0 fe (theory)	2.55	0.12	0.05	2.67	O.Hjortstam et al. [61]
3.0 fe (clean)	2.29	0.24	0.10	2.53	T.Nakagawa et al. [97]
3.0 fe + 0.1 K	1.91	0.17	0.09	2.08	T.Nakagawa et al. [97]
3.4 fe (clean)	3.33	0.23	0.07	3.56	J.Hunterdunn et al. [66]
3.4 fe (clean)	2.80	0.18	0.06	2.98	D.Schmitz et al. [117]
8.0 fe (clean)	0.80	0.08	0.10	0.88	D.Schmitz et al. [117]

Table 5.1: Magnetic moments of Fe/Cu(100). Concerning the errors for data from this work: For the thin film preparation it was ± 0.2 ML, while the relative error between the two sets was below 0.1 ML. The magnetic moments are extracted with the XMCD sum rules and therefore have typical error of ±5% for the ml to ms ratio and ± 10% for the seperated moments.

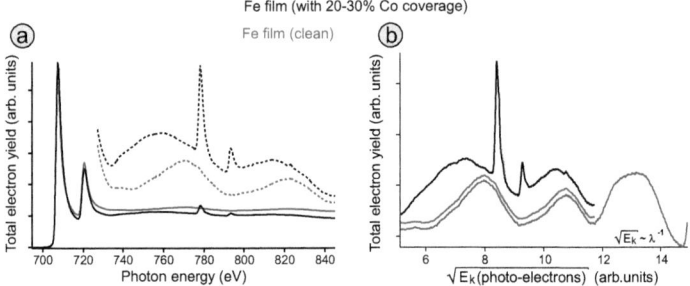

Figure 5.3: Fe EXAFS oscillations displayed as measured vs photon energy (a) and vs the square root of the kinetic energy of the photo electrons (b), zero being the Fe-L_3-edge. The EXAFS oscillations should occur regularly in a plot as in (b). One can qualitatively see that there is some change from the clean Fe surface spectra (red lines) to the 20% to 30% Co covered spectrum (black line), but for qualitative analysis, more detailed investigation in a wider energy rage is necessary.

of the third degree in the two missing sections that contain the Co absorption features. This is shown for the measured background in figure 5.4 and for the cluster spectra in figure 5.7.

By removing the background in the above described way, only the spectrum containing the actual $2p$ to $3d$ transitions remained. By removing a fitted background, the usually used double step function is replaced by a diagonal cutoff of the L-edges. Figure 5.5 shows a spectrum of a very slightly oxidized chemically prepared CoPt nanoparticle (section 4) and the usually applied step function, as well as the cutoff version used for the mass selected clusters. This procedure changes the line shape of the resulting white line spectrum slightly, but the line shape has no influence on the sum rules. When integrating the area under the curve as done for the sum rules, the difference between the two methods used is only 1%, which is a

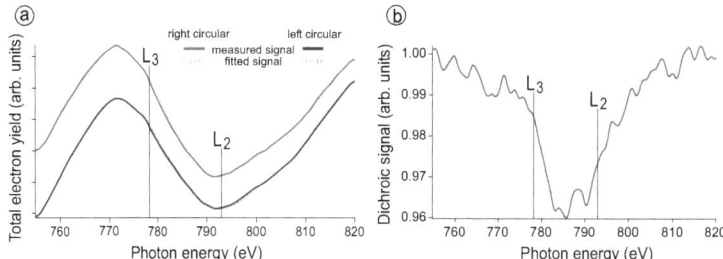

Figure 5.4: (a): There is an iron EXAFS oscillation, but no structure located specifically at the cobalt L-edges, that therefore could be related to any cobalt contamination of the beamline. Hence the background can be successfully approximated with a smooth fit that imitates the iron EXAFS oscillation. (b): The dichroic signal of the background measurement shows that there are iron MEX-AFS oscillations with a minimum right between the two cobalt L-edges. This makes it necessary to remove the iron MEXAFS oscillations from the cluster dichroism spectra, before applying the sum rules.

rather small additional error, when comparing that to the error of the sum rules (section 2.1.1).

The magnetization of each film was extracted from the data as in the case of nickel with the XMCD sum rules, but was not used to normalize the magnetic moments of the deposited clusters. The iron film magnetization changes last at the surface layer. The cluster magnetization couples by exchange coupling to the topmost layer of the magnetic substrate and hence the magnetic moments of the clusters have not to be normalized to the magnetic moment of the underlying magnetic thin iron film. This is different for nickel thin films, since there the spin reorientation transition from out of plane to in plane takes place in the growing and enlarging of magnetically differently orientated islands and thus the cluster on such islands have different magnetic orientation, thus not contributing to the

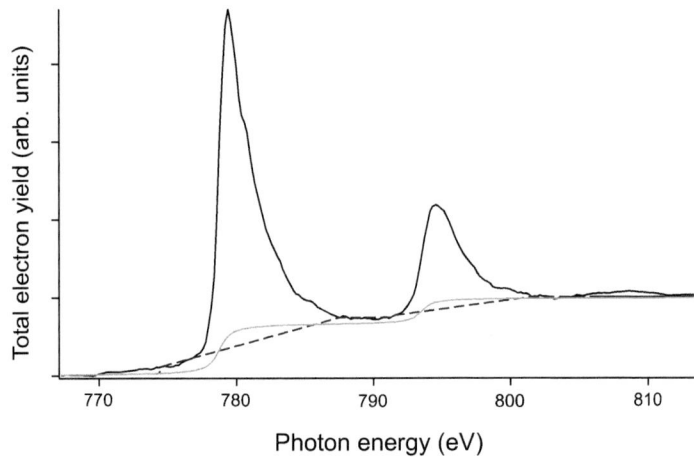

Figure 5.5: Comparing different background subtraction methods: For the clusters on the iron substrate it is impossible to first remove the background and later subtract the step-function properly. By approximating the background with a fit the step-function is in first order included into the background. Instead off removing a proper step-function (orange curve) by subtracting the approximated background a cut off is removed from the spectrum similar as shown in the picture (dotted line). This obviously changes the line shape of the resulting spectrum, but the area under the curve for the separate L-edges is affected less than 1% for Co. Since the XMCD sum rules do not depend on the line shape, this approximation method adds an additional error of about 1% only.

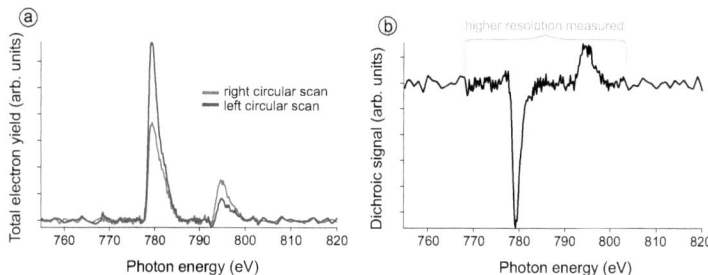

Figure 5.6: Removing the iron EXAFS and MEXAFS components from the measured cluster spectra the resulting spectra for the application of the sum rules are shown above.

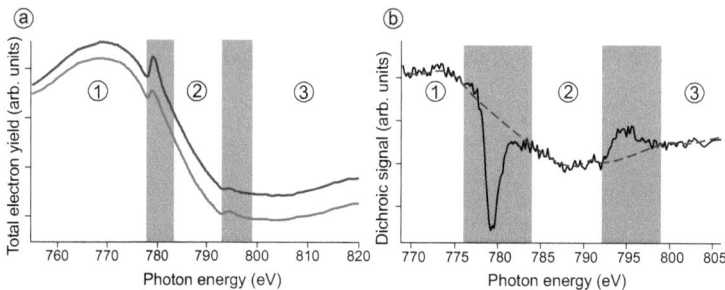

Figure 5.7: The spectra were fitted in the mask regions (1-3) with polynomial functions of 9th degree and mended in the intermediate regions with a polynomial of the 3rd degree. These merged fit functions were used as iron background.

perpendicular orientated XMCD measurement.

As a measure of the film thickness the step height to background ratio of the L-edges of iron was taken (figure 5.8). This works well because the iron film was very thin (≈3 layers) and the deviation from the average thickness was only half an atomic layer. In the small range of ± 0.5 layers in which the films could still be magnetized perpendicular, the absorption signal is in good approximation linear to the coverage. Additionally the background of an iron film on copper is featureless in the pre L_3-edge region and neither iron, nor cobalt or platinum has absorption features in this energy. The pre-edge region was normalized to one, hence the step height was directly accessible. A relative magnetization was defined by the magnetization measured divided by the over all maximal magnetization measured during the beamtimes. The relative magnetization versus this thickness measure is shown in figure 5.9a and shows a similar behavior of magnetization vs film thickness as the reference found in literature (figure 5.9b).

Spectra used in this work had a magnetic substrate with at least 75% relative magnetization. By this restriction the iron thin films used for the cluster measurements differ less than 0.5 ML in thickness offering similar surface conditions for the different cluster samples.

5.2 XAS whiteline spectra

Some clusters possess highly enhanced chemical reactivity, which is one of the reasons for the experiments in this work to be performed in UHV conditions. The L_3-edge line shape of the absorption signal of metallic and oxidized transition metals as cobalt differ strongly in

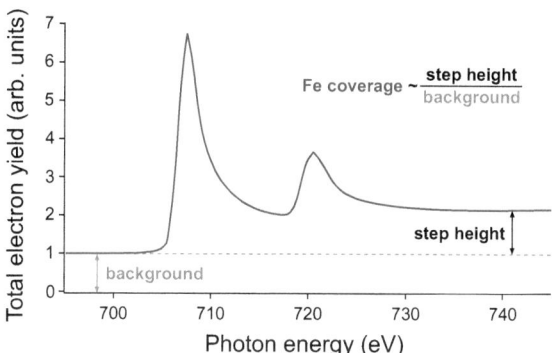

Figure 5.8: The iron films were deposited in thicknesses close to 3 atomic layers using a thermodesoption spectroscopically calibrated metal foil evaporator. Only in the small range of ± half a layer in which the films can be magnetized perpendicular to the substrate surface, the absorption signal is in good approximation linear to the coverage. The introduced measure is the step height of the L_3 edge jump divided by the background. This measure was used to monitor the grown film thickness. In the graph the background was normalized to one.

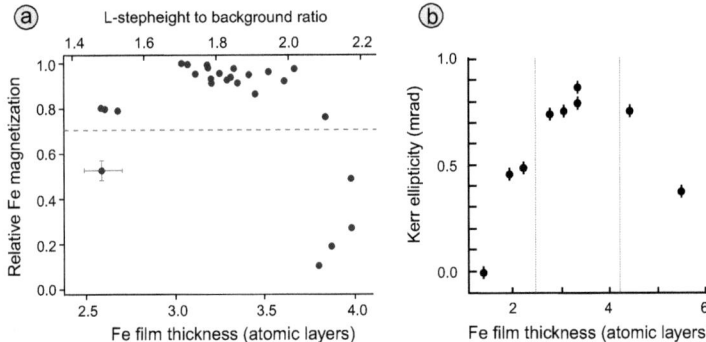

Figure 5.9: The iron film thickness has been estimated according to figure 5.8. By plotting this estimated thickness against the measured perpendicular magnetization (a) one can recalibrate the film thickness to atomic layers, since iron films on copper can only be perpendicular magnetized while the iron film is between 2.5 and 4.0 atomic layers thick. (b) shows a reference magnetization to thickness measurement of M.Kurahashi et al. [76], which was done with Kerr rotation measurements.

appearance (figure 4.3), which in the case of the wet chemical clusters discussed in the last chapter, helped to see that many preparations were strongly oxidized. In our measurements of deposited mass selected clusters the goal was to investigate the magnetic properties of unoxidized clusters. The Co_n and Co_nPt_m clusters on the iron thin film were with one exception all oxide free and showed just small effects due to aging over the 4 hour time span of the measurements. The Co_n clusters on nickel were all oxide free as well and did not age significantly. Even after dosing pure oxygen upon a Co_1 cluster preparation, only the magnetization of the iron thin film disappeared and with that the magnetization of the clusters too, but the cobalt atoms remained metallic. After annealing to 120K in the presence of oxygen, finally the Co atoms could be oxidized. In contrast to that it proved hard to prepare Co_2Pt_1 and impossible to stabilize Co_1Pt_1 on Ni/Cu(100). Using the measure to estimate the oxidation of a Co spectrum defined in the chapter for wetchemical CoPt nanoparticles (figure 4.2), all Co_1Pt_1 mass selected cluster preparation were at 50% oxidation already at the first scan and averagely oxidized up to 65% within 3 hours. The comparison with spectra taken of oxidized wet chemical nanoparticles is shown in figure 5.10. The presence of Pt on the nickel substrate has an enormous catalytic effect upon the oxidation process of the cobalt atoms, which implies that the platinum must lead to an increased electron density at the cobalt atom. This effect would be lessened if one Pt atom hybridizes with two Co atoms as in Co_2Pt_1, in which case the increased reactivity is still present, but less pronounced. The energy position of the cobalt edges shift between the deposition upon iron and nickel substrates for all clusters (figure 5.12a). It is

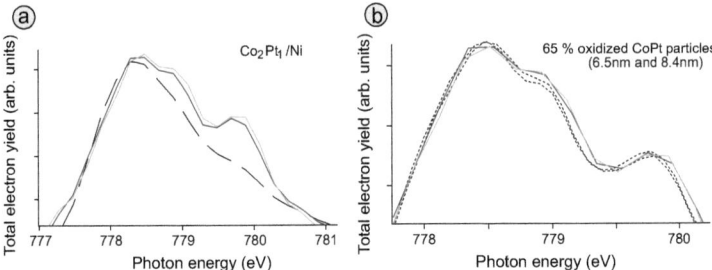

Figure 5.10: There were strong signs of oxidation visual in the absorption spectra of some preparations (a). The almost identical samples are oxidized Co$_2$Pt$_1$ clusters on nickel, whereas the dashed line is a not oxidized preparation of Co$_2$Pt$_1$. The oxidized cobalt in the cluster appeared only on nickel surfaces and when platinum was present. Comparing the two spectra of oxidized mass selected clusters from (a) with some of the spectra of the 65%oxidized wet chemically prepared particles (a 6.5nm and a 8.4nm nanoparticle), it shows that the line shape of all 4 spectra are almost identical.

roughly 0.5 eV for all clusters measured. The exchange interaction between iron and cobalt is much higher than that of cobalt and nickel. This can lead to a shift of the electron density from cobalt to iron in case of the neutral cobalt ad-atoms, which would leave the cobalt d-orbitals less populated in the initial state and shift the L$_3$-edges towards lower energies. On the other hand the strong coupling to the iron surface will allow the electrons to help shield the produced core holes in the final state. A difference in the shielding of the final state core hole can be seen, when comparing the Co$_1$ and Co$_1$Pt$_x$ clusters on iron, in which the Co-Pt $3d$-$5d$ hybridization effectively increases the electron density at the cobalt atom in the final state and thus increases the BR by 50% (figure 5.11). This would explain the increased reactivity of Co$_1$Pt$_1$ on nickel. On both surfaces the Pt would donate some electron density by $3d$-$5d$ hybridization. On the nickel surface this additional electron density is almost completely at the location of the Co atom, which will increase the attractive force towards oxygen. On the iron surface the platinum hybridizes strongly with the iron substrate too, hence the charge at the location of the cobalt atom is almost unchanged.

In our previous experiments with chromium clusters on iron [111], free cobalt clusters [82] or as it has been reported for Co atoms and thin films on a copper surface [31] changes of the energy position of the cluster L-edges could be seen that depended on the cluster size or chemical surrounding of the deposited material. Principally the energy position of the absorption edges react to any changes made to the system investigated usually explained with initial and final state effect. M.Reif et al. explained the shift of the Chromium absorption edges with increasing cluster size as an initial state effect of increasing $3d$ electron delocalization due to intra cluster hybridization and a corresponding lowered $2p$ binding energy. Since the clusters investigated in this work were very small, up to Co_3 and $CoPt_3$, further investigations with larger clusters may be more conclusive. The influence of the Pt in the cluster on the Co core level splitting or the influence of the substrate can not be satisfyingly deduced from the data measured so far.

The changes of the energy positions of the $L_{2,3}$-edges are much smaller for clusters on the stronger coupling iron thin film than for those on the nickel surface. Likewise the measured apparent spin-orbit splitting changes less for samples on the iron thin film. The effect of the platinum in the clusters is little, although it appears that the platinum shifts the absorption edges of the clusters towards lower binding energies on both substrates.

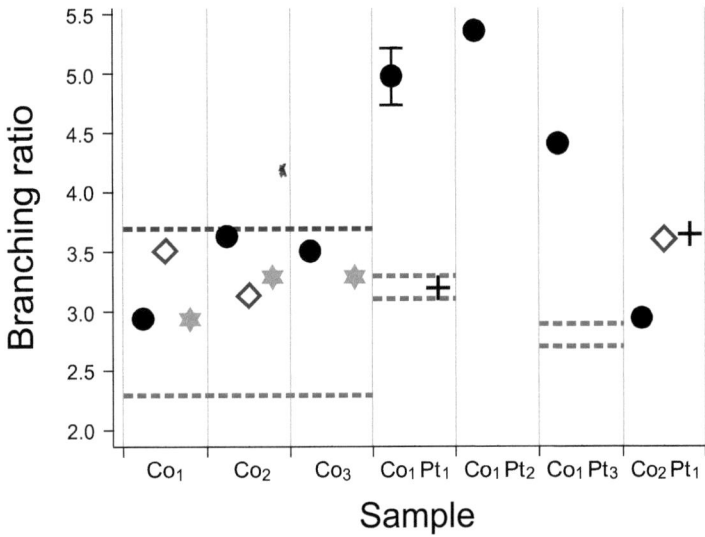

Figure 5.11: The ratio of the integrated absorption over the cobalt L_3 edge divided by the integrated absorption over the L_2 edge show that there are effects that seem to depend on the surface and on the oxidation state of the cobalt. The black solid circles are the experimental values from this work on the iron substrate, while the blue open diamonds are those on the nickel substrate. The black crosses are oxidized cluster preparations of this work. The orange stars are from free cluster experiments by J.T.Lau et al [82], the green dotted line (value 2.3) at Co_1 to Co_3 is the BR of bulk cobalt [9], the brown dotted line that of CoO nanoparticles [69]. The green dotted lines at Co_1Pt_1 and Co_1Pt_3 are the thin film values taken from [24, 52, 53, 68]. The oxidized particles behave very similar no matter whether pure cobalt or an CoPt alloy is present, while the branching ratio of the not oxidized clusters increases strongly in the presence of platinum.

144

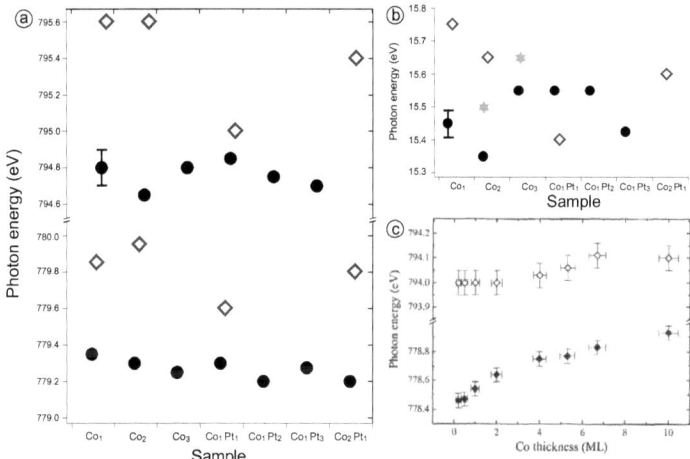

Figure 5.12: The maximum position of the Co L-edges for the different clusters is displayed (a) and the resulting Spin orbit split that is measured (b). The solid black circles represent cluster measurements on Fe/Cu(100), while the blue diamonds are clusters on Ni/Cu(100). In comparison the position of the L-edges of a thin Co film on Cu(1 1 13) measured by H.Dürr et al. [31] is displayed in (c) and there are two data points added to (b), that were extracted from free Co cluster experiments published by J.T.Lau et al. [82] and are displayed as orange stars.

The shift Dürr et al. measured from sub-monolayer to 10 ML thick films (all deposited on Cu(1 1 13)) was less than 0.1 eV at the L_2-edge and about 0.5 eV at the L_3-edge. They explained the change with the increase of the coordination of the cobalt atoms and the general change of atomic surrounding and could identify that the lower energetic (surface) peak remains as a shoulder in the bulk like absorption line of thicker coverage. The lowest coverage of Dürr et al. was in the sub monolayer regime with a $L_{2,3}$-edge splitting of 15.5 eV. This is very in good agreement with our measurements that range from 15.8 eV to 15.4 eV. The measured SO splitting reduces for Co_1 clusters on iron from 15.5 eV to 15.4 eV, when increasing the coverage from 3% to 25%. This can be explained as in the case of the thin films investigated by H.Dürr et al.: At a coverage of 3% almost all small Co clusters are separate where as in the case of 25% coverage most clusters are joined to islands. The coordination of the Co-atoms change and especially the atoms with 3 or more Co neighbors increase strongly. In the study of H.Dürr et al. the value of 15.4 eV for the $L_{2,3}$-edge splitting is reached at a coverage of one full layer. Considering the strongly stepped Cu(1 1 13) surface used for that study, it is most probable that the increase of Co coverage leads to stripes along the edges of the Cu crystal and not to islands. This would also explain, why there is almost no change in the data of H.Dürr et al. below 1 ML Co coverage and rather drastic changes between 1 and 4 layers Co coverage. It looks different with a flat Cu(100) surface with pseudomorphic grown iron or nickel thin films as in the case of this work. Here it is most likely that the formation of islands starts early and long 2 dimensional lines are rather scarce. One will expect features due to coordination effect to appear at lower coverage, since higher coordinated ad-atoms are more likely on flat surfaces.

5.3 XAS dichroism spectra

The evaluation of the orbital and spin magnetic moments of the deposited Co_n and Co_nPt_m clusters can be done using the sum rules (equations 2.3 and 2.4). For a general overview and the trend of the orbital to spin magnetic moments one can just look at the dichroism spectra that are normalized to the L_2-edge:

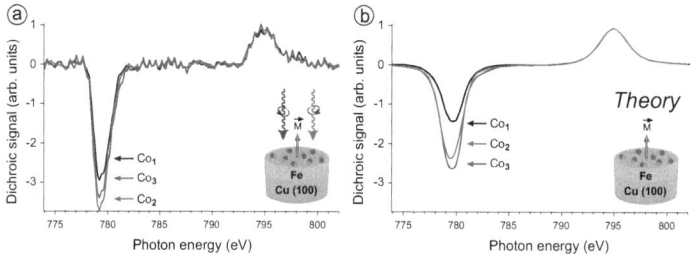

Figure 5.13: The experimental dichroism spectra of Co monomer, dimer and trimer on Fe/Cu(100) normalized to the integrated intensity of the L_2-edge (a). The system was calculated with a full potential Korringa Kohn Rostoker Green functional method (b) and the result comes close, but still there are distinct differences between theory and experiment.

The rather simple system of Co_1 to Co_3 on iron was calculated as a comparison to the experiment with a full potential Korringa Kohn Rostoker Green functional method (SPR-KKR) by S.Bornemann [19] (figure 5.13b). The trend of the theoretical calculations is different to the experiment, since the theory predicts a monotonous increase of the orbital to spin moment ratio, while the experiment is showing an increase from monomer to dimer and a decrease from dimer to trimer. One reason for the differences can be that the theory at the moment always assumes flat surface layers, which in the case of the Fe/Cu(100) system is a rude approximation. Furthermore strongly differing from the experimental conditions the iron was assumed to grow in bcc-Fe(100) configuration. At last the theoretical clusters were calculated as sitting on top of the surface: The clusters in the experiment were deposited on a rough surface and are likely to be distributed between on-top positions and some that are at steps or in the first surface layer. Calculations of S. Bornemann [19] that simulated a ruthenium monomer sinking into an iron thin film, shown that the magnetic moments differ strongly between those positions and even invert their sign, when sinking from a on-top position into the topmost film layer. After soon finished updates of the SPR-KKR code some detailed calculations concerning these effects will be done. The theoretical evaluated orbital to spin ratios give smaller values than the evaluations using the sum rules, when calculating spectra from theoretical SPR-KKR calculations and evaluating the dichroism with the sum rules, the ratio of orbital to spin moment increases 20-30%. Recent yet unpublished calculations of A.Lichtenstein [83] of cobalt monomers and dimers on the Fe/Cu(100) surface with LSDA+U (including spin correlation effects) have given higher orbital to spin moments than the experimental data (table 5.5). Although it is tempting to accept the achieved congruities of theory and experiment one needs to remind that in the calculations the surfaces were flat (in the case of A.Lichtensteins calculations the surfaces are rather small island) and the clusters did not sit on different positions as described in the case of the ruthenium cluster calculations of S.Bornemann. Theory not including electron correlation effects underestimates the orbital moment in comparison to the spin moment considerably, but can be used to compare trends. Further characterization of the thin film systems used is highly important to effectively compare any measurements with theoretical calculations. There is no energetic shift measurable in the experiment from the cobalt monomer to the cobalt dimer, as can be seen in the SPR-KKR calculation (figure 5.13b). There are yet no calculation for the nickel films performed, although the approximation of a flat surface is better to be used for Ni/Cu(100) than for Fe/Cu(100). It would be interesting to see whether the theory can reproduce the increase of the orbital to spin moments of cobalt from iron (figure 5.13a) to nickel (figure 5.14) that in the experiment is almost a factor of two. A simple idea to explain this is that the coupling between iron and cobalt is much stronger than between cobalt and nickel and thus the clusters are more free on nickel and the orbital moments are quenched due to the strong coupling by the iron film. This idea is supported by the very high orbital to spin magnetic moments that were measured by P. Gambardella et al. [50] for cobalt monomers on Pt(111), where the Co atoms couple only very weakly to the Pt surface. Comparing the orbital moments of the Co clusters on iron and nickel (table 5.2 and table 5.4) one can see that the orbital moments are almost identical on both surfaces, but the spin magnetic moment of Co clusters on the iron substrate are 20-30% higher than on a nickel surface. The strong coupling of the iron thin film therefore seems not to quench the orbital moment, but to boost the spin moment. Maybe it is a little of both, since all experimental data are taken without the knowledge of the number of d-holes of the Co clusters. If the iron thin film would quench the total orbital moments of the Co clusters, while decreasing the number of empty d-states in the Co clusters at the same time.

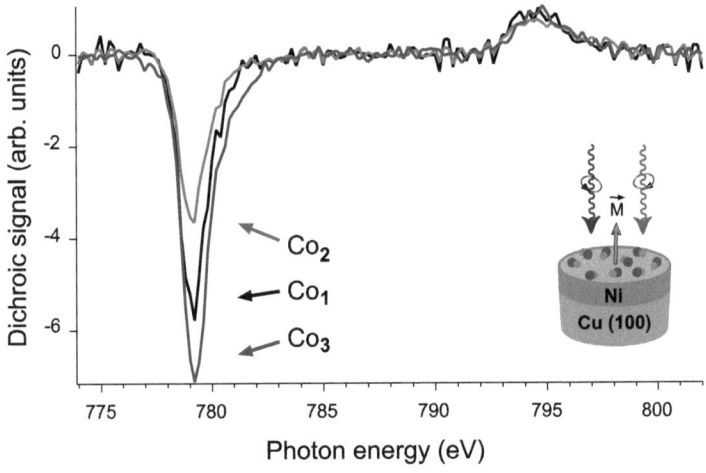

Figure 5.14: Co monomer and dimer with and without Pt on Ni/Cu(100) normalized to the integrated intensity of the L_2-edge

To investigate the effect of adding platinum to the Co clusters there Co_1 was deposited on an iron film. Larger clusters with increasing number of platinum atom up to Co_1Pt_3 (figure 5.15) were successively prepared. Adding one or two platinum atoms increased the orbital to spin moment ratio, but the third brings it back to the starting value. Overall the changes of $m_l/m_{s(eff)}$ are rather small and almost within the uncertainty of the measurements of about 10%. This rather small effect can be again due to the strong coupling of the cobalt to the iron thin film.

The effect of adding a platinum atom to a cobalt dimer is very large. The orbital moment of the cobalt doubles (figure 5.16 and table 5.4).

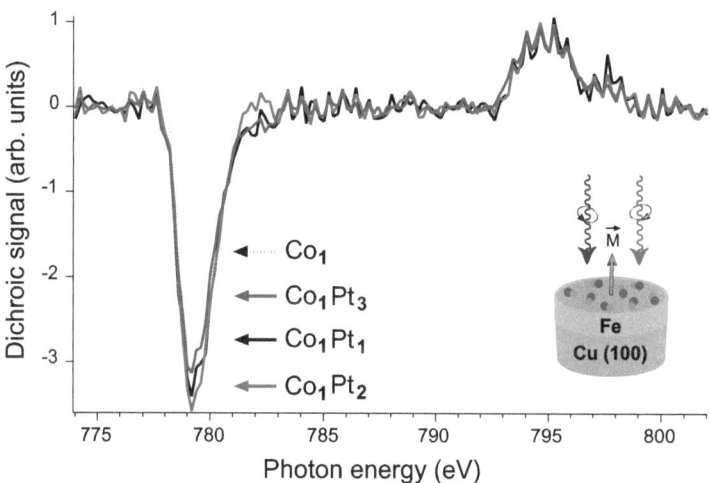

Figure 5.15: Comparison of the dichroism spectra of Co monomer and after adding one two or three Pt atoms to the cluster on Fe/Cu(100) normalized to the integrated intensity of the L_2-edge

The comparison with the iron thin film system is unfortunately impossible yet, since the dimer was measured on iron several times, but not in sufficient quality [1]. The Co_2Pt_1 on Fe/Cu(100) was evaluated and is shown in the last line of table 5.2. Obviously they can not be taken as a reference. The other possible candidate for a direct comparison of CoPt clusters on iron and nickel surfaces would have been the Co_1Pt_1. This cluster was easy to prepare on an iron surface, but was the only cluster to oxidize at every preparation on the nickel surface. This is on the one hand not highly unlikely, since cluster are much more reactive than the bulk material, although this is rather not the case for cobalt. None of the cluster preparations of CoPt-clusters on the Fe/Cu(100) surface oxidized and none of the pure cobalt clusters on the Ni/Cu(100) surface oxidized. Even after adding pure oxygen heating was needed to achieve oxidization of pure cobalt clusters on Fe/Cu(100). All 4 preparations of Co_1Pt_1 on Ni/Cu(100) were oxidized, which points strongly to a highly catalytic effect of platinum in this configuration. Additionally the ratio of orbital to spin moment in the oxidized Co_1Pt_1 cluster on Ni/Cu(100) was lower than for any not oxidized cobalt cluster.

The magnetic moments of the two measured oxidized cluster types behaved mainly as expected, since in oxides both spin and orbital magnetic moment decrease. The samples had an estimated oxidation level of 65% and the spin moment of both clusters was at less than 50% of any of the not oxidized. The drop of the orbital moment was as well very strong, but only in the case of the Co_2Pt_1

---

1. The synchrotron had a quite unstable beam position during those measurements, leading to slight variations of the focus spot of the light, which leads to measuring different positions of the gaussian distributed cluster spot, making the measured tey proportional to the beam position, with an unknown proportionality function.

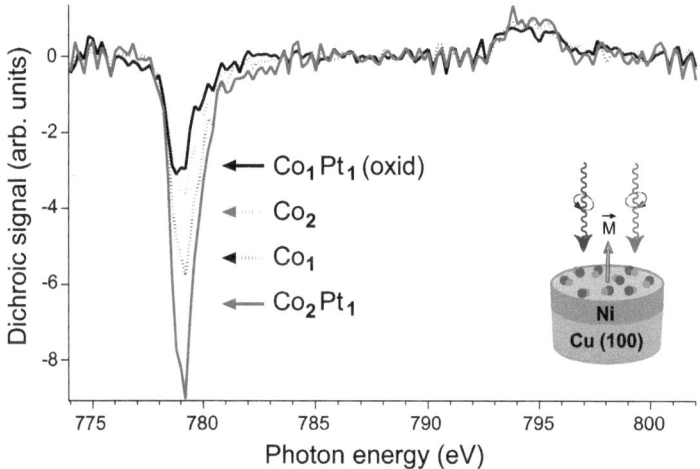

Figure 5.16: Comparison of the dichroism spectra of Co monomer and dimer on Ni/Cu(100) and adding a Pt to either cluster. All is normalized to the integrated intensity of the L_2-edge

the orbital to spin ratio was really high, although it had been like that for the not oxidized clusters too. Typically this ratio increases

sample	$m_s/(\mu_B \cdot d_h)$	$m_l/(\mu_B \cdot d_h)$	m_l/m_s	$mag_{tot}/(\mu_B \cdot d_h)$
Co_1	0.61	0.10	0.16	0.71
Co_2	0.66	0.15	0.22	0.81
Co_3	0.76	0.17	0.22	0.93
Co_1Pt_1	0.69	0.15	0.22	0.84
Co_1Pt_2	0.65	0.13	0.20	0.77
Co_1Pt_3	0.71	0.14	0.20	0.85
Co_2Pt_1	0.88	0.05	0.06	0.69

Table 5.2: Magnetic moments of CoPt clusters on Fe/Cu(100). The effects are all together small, while the last line shows the evaluated data of highly problematic measurements, which results are certainly not reliably.

sample	$m_s/(\mu_B \cdot d_h)$	$m_l/(\mu_B \cdot d_h)$	m_l/m_s	$mag_{tot}/(\mu_B \cdot d_h)$
Co_1	0.51	0.15	0.30	0.66
Co_2	0.57	0.12	0.20	0.69
Co_3	0.45	0.16	0.35	0.61
Co_2Pt_1	0.62	0.24	0.39	0.86

Table 5.3: Magnetic moments of CoPt clusters on Ni/Cu(100). There is a huge effect, when adding Pt to a Co dimer, the orbital moment doubles.

sample	$m_s/(\mu_B \cdot d_h)$	$m_l/(\mu_B \cdot d_h)$	m_l/m_s	$mag_{tot}/(\mu_B \cdot d_h)$
Co_1Pt_1	0.21	0.04	0.17	0.25
Co_2Pt_1	0.24	0.09	0.38	0.33

Table 5.4: Magnetic moments on nickel

when changing from cobalt to cobalt oxide roughly 50%. This time it remained almost constant for Co_2Pt_1 and did probably even drop for Co_1Pt_1. We could not measure the magnetic moments of not oxidized Co_1Pt_1 on nickel, since the earlier described very high oxidation speed of those clusters. All measured clusters on nickel had a higher m_l/m_s ratio than 0.17, and thus it is most likely that Co_1Pt_1 would have too. The doubling of the orbital moment from Co_2 to Co_2Pt_1 makes it rather hard to imagine that there should be a decrease from Co_1 to Co_1Pt_1. Possibly the Oxide binds at different position of the two clusters with respect to the surface (parallel to the surface for Co_1Pt_1 and on top for Co_2Pt_1). If the Oxygen should prefer a special geometry it could induce differently spacial orientated orbital's at the cobalt. Since all the cluster measurements were performed in normal incidence, a tilt of the orbital moment into the films plain would appear as a reduced orbital moment, as maybe in the case of the oxidized Co_1Pt_1. Measurement of CoO clusters on a nickel film are planned and might give some inside how oxide clusters orientate on the surface. On the other hand the clean Co_2Pt_1 has shown a very high orbital magnetic moment and since only 65% of the clusters were oxidized, the remaining clusters might be responsible for this high, but not increased out of plane orbital to spin ratio.

Comparing the orbital to spin moment ratios of Co_1 on Fe and Ni and Pt(111) as approximately done in figure 5.17, it is apparent that the ratio almost doubles from Fe to Ni and than more than triples from the Ni to the Pt substrate. This points towards an increasing orbital to spin ratio, when decreasing the cluster substrate interaction. Above that Pt seams to have an strong increasing effect

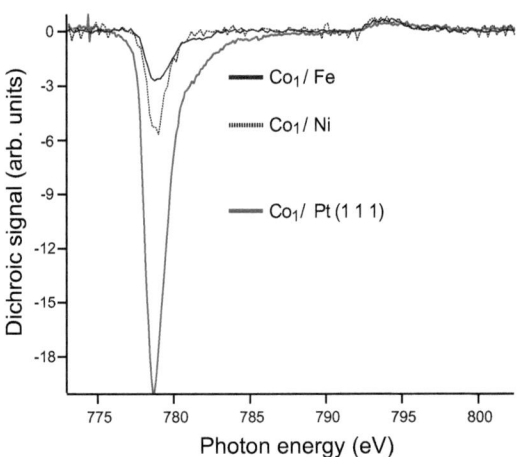

Figure 5.17: A comparison of the orbital to spin ratio of atoms on three differently strong coupling substrates shows that the ratio increases, when the substrate atom coupling weakens. The Pt(111) surface reference was measured at 5.5 K and 7 T external field by Gambardella et al. [50]

on the orbital moment of the clusters, when the cluster substrate interaction is weak. In the case of the strong coupling Fe substrate, the enhancing effect on the Co orbital moment of the Pt appears to be quenched. Taking in account that the orbital moment of Co_2 on Ni doubles when adding Pt the strong impact of the Pt surface to the orbital to spin moment ratio of the Co may not only be due to weaker substrate cluster coupling, but due to enhancement effects upon the orbital moment by $Co3d$-$Pt5d$ hybridization. Please note that there are differences in the size dependence upon the different substrates. While the measurements of Gambardella et al. on the Pt surface have shown monotonous decrease of the orbital to spin moment ratio with increasing cluster size, this ratio decreases from Co_1 to Co_2 and is largest for Co_3 on Ni and again different for the Fe substrate, where the ratio increases from Co_1 to Co_2 and has an intermediate value for Co_3.

Taking a closer look at the magnetic moments of Co_nPt_m separately comparing again the Fe and Ni substrate influences (table 5.2 and 5.4). It seems that the Co spin moments are all together larger on the Fe substrate. As described in section 2.1.1 the measured spin moment with XMCD is only an effective spin moment, including seven times the z component of the magnetic dipole operator: T_z. In spherical geometry T_z is negligible and is typically assumed to be zero when unknown. For Fe, Co and Ni bulk and thin film systems all experimental data have so far supported this practice, but as discussed in chapter 4 the influence of the strong spin orbit coupling of the Pt may lead to a non diminishing component of T_z, although it was not the case for monoatomic Co chains on Pt(997). In the case of Co_n on an Fe substrate the coupling is obviously strong enough to quench almost all effects due to Pt ad-atoms. With this strongly interacting iron underneath and vacuum above the coupling to the Co is very different on two opposing sides and hence it is to be expected that the Co has no spherical surrounding, giving rise to a none zero T_z. To the best of the knowledge there are no fully relativistic theoretical calculations including electron correlation effects of the value of T_z for Co_n on Fe yet. To estimate the magnitude of T_z for Co_n on the Fe substrate and one can use the weakly coupling Ni substrate as reference and assume the T_z to be zero there. This assumption is feasible, especially when comparing the mass selected measurements with the measurements of the wet chemically prepared nanoparticles (chapter 4) and the reference Co nanoparticles [12, 13, 73]. The spin moment per d-hole of Co of those pure Co nanoparticles are very similar to those of the mass selected Co_n and Co_nPt_m clusters on a Ni substrate. Lets assume the spin moments of Co per d-hole to be approximately equal on iron and on nickel substrates. Averaging over the Co_1 and Co_2 measurements and taking the difference of the measurements on iron and on nickel as an estimate of the $7 \cdot T_z(Fe)$ contribution in the XMCD measurement of the Co spin moment $T_z(Fe) \approx 0.06$ μ_B (using $n_{d-holes} = 2.9$ for Co on iron [83]). As a result of this estimation 35% of the XMCD measured Co spin moment on the iron substrate would originate from T_z. The uncertainty of the measured moments would only in case of the orbital moment be limited by the measuring method (XMCD + sum rules), while the spin moment would be dominated by the uncertainty of T_z. It is essential that by further theoretical and experimental work uncertainties concerning T_z are strongly reduced. Nevertheless the trends described above stay true, since the Co_n and Co_nPt_m spin moments within the measured data set for each substrate vary less than half as much as the spin moments of equal cluster size vary comparing different substrates. Therefore the T_z seems to be rather similar for all clusters on a certain substrate and the trends within a measured data set upon a substrate remain true.

sample	$n_{d-holes}$	$\frac{m_s}{\mu_B \cdot d_h}$	$\frac{m_l}{\mu_B \cdot d_h}$	$\frac{m_l}{m_s}$	reference
$Co_{(23layer)}$	2.8	0.982	0.156	0.16	[67]
$Co_{(bulk)}$		$1.52/d_h$	$0.14/d_h$	0.09	[38]
$Co_{(bulk)}$		$1.607/d_h$	$0.155/d_h$	0.10	[145]
$Co_{(layers-fcc)}$				0.110	[98]
$Co_{(layers-hcp)}$				0.148	[98]
$Co_{(2-10ly)}/Pt(111)$				0.2	[134]
Co_{300}/Au				0.21	[30]
$CoPt_{(3nm)}$ [as prep.]	2.628	0.647	0.046	0.071	[141]
$CoPt_{(3nm)}$ [annealed]	2.628	0.727	0.068	0.094	[141]
$CoPt_{(bulk)}$				0.15	[145]
$CoPt_{(40nm-film)}$	2.628	0.75	0.099	0.13	[52]
$Co/Pt_{(multilayer)}$	2.49	0.627	0.052	0.08	[24]
$CoPt_{3(film)}$	2.25	0.711	0.058	0.19	[53]
$CoPt_{3(bulk)}$				0.26	[68]
$CoPt_{3(bulk)}$				0.094	[145]
$Co_{n(n=20-200)}$		m_{tot}=2.08±0.20μ_B			[23]

Table 5.5: Literature references of experimental values of magnetic moments of Co and CoPt clusters, layers and bulk systems

sample	$n_{d-holes}$	$\frac{m_s}{\mu_B \cdot d_h}$	$\frac{m_l}{\mu_B \cdot d_h}$	$\frac{m_l}{m_s}$	reference
Co_1/Fe(100) [LSDA+U]	2.91	0.701	0.172	0.245	[83]
Co_2/Fe(100) [LSDA+U]	2.90	0.679	0.207	0.305	[83]
$Co_{(wire)}$/Pt(997) [LMTO+OPT]		$2.09/d_h$	$0.86/d_h$	0.41	[75]
$Co_{(bulk)}$ [LSDA+DMFT]		$1.614/d_h$	$0.138/d_h$	0.09	[145]
$CoPt_{(bulk)}$ [LSDA+DMFT]				0.139	[145]
$CoPt_{(bulk)}$ [FP-LMTO LSDA]	2.628	0.681	0.041	0.061	[49]
$CoPt_{(bulk)}$ [FP-LMTO GGA]	2.628	0.696	0.033	0.048	[49]
$CoPt_{3(bulk)}$ [LSDA+DMFT]				0.099	[145]
$CoPt_{3(bulk)}$ [FP-LMTO LSDA]	2.651	0.694	0.018	0.03	[49]
$CoPt_{3(bulk)}$ [FP-LMTO GGA]	2.651	0.720	0.022	0.03	[49]
Co_{13}	2.2	$m_{tot}=2.02\mu_B$			[1]
$CoPt_{(cluster)}$		$m_{tot}=2.75\mu_B$			[44]
$(CoPt)_{2(cluster)}$		$m_{tot}=1.99\mu_B$		0.15	[44]
$Co_{(bulk-fcc)}$		$m_{tot}=1.64\mu_B$			[95]
$Co_{(bulk-bcc)}$		$m_{tot}=1.73\mu_B$			[95]
$Co_{(bulk-hcp)}$		$m_{tot}=1.63\mu_B$			[95]

Table 5.6: Literature references of theoretical values of magnetic moments of Co and CoPt systems

Chapter 6

Summary

6.1 Conclusion

The goal of this work was to increase the experimental understanding of size and stoichiometrically induced property changes of small magnetic clusters and nanoparticles using x-ray spectroscopic methods. In this context first x-ray absorption spectroscopy (XAS) and x-ray magnetic circular dichroism (XMCD) measurements of deposited size selected Co_nPt_m clusters (n,m $\leq$ 3) and of wet chemically prepared Co_xPt_{100-x} nanoparticles (3.7 nm to 8.4 nm) could be presented in this work. The chemically selective measurements of XAS and XMCD were well suited to characterize these systems concerning their oxidation state and their magnetic properties.

The small mass selected clusters have proven as expected to be excellent systems to prepare well defined samples and they can to some extend be calculated with high sophisticated ab-initio theory methods. The clusters were magnetically aligned perpendicular to the surface using exchange coupling to the upper layer of Fe/Cu(100) and Ni/Cu(100) thin film substrates. For theory free clusters are generally easier to calculate, but to bridge the gap from pure model systems to real samples, including the cluster substrate interaction is essential. The rapid oxidation of most deposited clusters consisting out of very few atoms require in situ preparation of samples at ultra high vacuum (UHV) conditions in this case with a base pressure of about $1 \cdot 10^{-10}$ mbar. Under these UHV conditions almost all Co and Co_nPt_m have proven very stable on both magnetic substrates within the usual measuring period of 4-5 hours. Only Co_1Pt_1 on a Ni substrate was always at least 60% oxidized within minutes. Additionally to this catalytic effect of the Ni substrate on the Co_1Pt_1 cluster, strong size and stoichiometry dependencies could be seen, as the orbital moment of the Co dimer doubled, when one Pt was added to the cluster. On the more strongly coupling Fe layer this stoichiometry effect was effectively suppressed. There are strong size dependencies in the orbital moment for pure Co_n clusters (n $\leq$ 3) on Fe and Ni substrates, although with different trends. The coupling of substrate and cluster worked both ways, as the 3% Co coverage of the deposited clusters had no effects on the weakly coupling Ni substrate, while it lead to a shift of several eV towards lower energies in the extended x-ray absorption fine structure (EXAFS) oscillations for the Fe substrate.

All Co spin moments per d-hole are rather well within experimental and theoretical reference values from literature (≈ 0.6 μ_B), while the measured orbital moments are a factor of 2-3 above most experimental reference values and roughly a factor of 5 above all theoretical values, where the electron correlation effects are neglected. This could be explained by the fact that the experimental references are for larger particles (300 atoms per cluster and more), where the quenching of the orbital moment, common for bulk material, may already have started. There are two theoretical calculations of which one for to Co_1 and Co_2 on Fe/Cu(100) [83] includes all electron correlation effects and of which the second partially readds the electron correlation effects by an optical polarization term [75]. These two theoretical calculations come to orbital magnetic moments per Co atom of 0.17 to almost 0.3 μ_B per d-hole, which is even slightly above the experimental values of this work (0.10 to 0.24 μ_B per d-hole). Since the use of the XMCD sum rules slightly underestimates the orbital magnetic moments [127] theory and experiment are in good agreement here. It proves that the electron correlation effects play a major role for the orbital magnetic moments of small clusters and that including electron correlation effects in theoretical calculations for these systems is essential. A size independent strong enhancement of about 20% of Co spin moments extracted with the XMCD sum rules changing from the Ni substrate to the more strongly interacting Fe substrate, points towards a non diminishing z-component (T_z) of the magnetic dipole operator.

Besides the fundamental research part of this work dealt with wet chemically prepared CoPt nanoparticles of 3.7 nm to 8.4 nm diameter with a sharp size distribution of only 10%. To characterize the particles with x-ray spectroscopic methods and measure their magnetic properties in their native state, the particles were prepared with their organic ligand shell as single layer films on p-doped Si wafers. The ligands play a major role in keeping the particles from agglomerating. The layer growth and the total conductivity was checked with scanning electron microscopy (SEM) prior to the x-ray spectroscopic measurements. In contrast to High resolution transition electron microscopic (HRTEM) measurements done right after synthesis of the particles, photo electron spectroscopic (PES) measurements [88] had given clear evidence of oxidized Co in samples of earlier synthesis batches. Using the x-ray absorption spectroscopy (XAS) measurement data, taking advantage of it's element specificity and chemical sensitivity a core-shell oxidation model was created including the size dependent stoichiometry changes that had been confirmed by the chemists. The stoichiometry changes from $CoPt_3$ for small particles to CoPt for large particles. All particles are already partially oxidized during synthesis at their surface due to the bonding of most outer Co atoms to the organic ligands. In contact to air the particles age over a period of roughly 6 months, while an oxide layer is formed up to 1 nm inwards, oxidizing only Co atoms that are not in stable $CoPt_3$ configurations. The oxide layer is less than half as thick than in equally sized pure Co nanoparticles [150], while additionally the oxidation speed is much slower changing the time scale from milliseconds to months.

The Co spin moments extracted from the XMCD measurements for all fresh prepared nanoparticles are about 0.55 μ_B per d-hole. This average spin moment is slightly below, while the average orbital moment of 0.12 μ_B per d-hole is slightly above literature values of equally sized unoxidized pure Co nanoparticles [12]. The reduced spin moment is explained with the partial oxidation of the particles and the fact that CoO is anti-ferromagnetic, thus not adding to the integrated signal of the XMCD measurement. The enhanced orbital moment compared with pure Co particles on the other hand was expected, since Pt was known to have an increasing effect on the Co orbital moments of neighboring atoms. No actual size dependence of the magnetic properties could be seen, which on the one hand is good for application purposes since it allows smooth scaling without any quantum size effects. On the other hand the particles possess not only a reduced volume with shrinking size, but additionally a reduced magnetic moment to volume ratio with decreasing particle size due to the decreasing relative Co content. This is not ideal if (as in this case) one is looking for particles which are most stable against thermally induced magnetization reversal, where with decreasing particle size it becomes more and more important that the magnetic moment per unit volume of the particle is as large as possible.

6.2 Outlook

For applicability purposes of the investigated wet chemically prepared CoPt nanoparticles the initial oxidation must be removed or avoided. With the used synthesis method avoiding the initial oxidation is not possible, but a ligand exchange after synthesis in favor of less strongly binding ligands or a complete removal of all organic ligands with oxygen and hydrogen plasma treatment may work. Additionally the oxidation due to aging must be prevented which could be done by adding a noble metal coating or by a comparable Aluminum oxide coating to keep oxygen out of the particle. An additional coating layer may as a bonus even shield the else harmful effects of Co on biological systems and might even open up a new field for a possible application of these particles.

Concerning the fundamental research: Unfortunately the calculation power of state of the art computation facilities is not big enough jet to calculate in a moderate time more than deposited $3d$ transition metal dimers. Larger and faster computation facilities may solve this limitation in the future. Until then theoretical calculations are urgently needed still, for already experimentally measured systems, as Co_n (n $\leq$ 3) on Ni/Cu(100). Some theoretical studies about size dependent changes of the empty d-states would help much to determine absolute magnetic moments with XMCD. Additionally an theoretical analysis of T_z for very small clusters should be done to reduce errors in the experimental analysis using the XMCD sum rules, starting possibly with Co_n (n $\leq$ 3) on Fe/Cu(100).

On the experimental side many different materials can be investigated in the future: Magnetic alloys similar to Co_nPt_m, like Co_nPd_m, Fe_nPt_m and other potential base materials for future magnetic application and to enhance the experimental hard proof to benchmark theory. Pure 3d metal oxide cluster that would help to understand the influence of oxidation and could give insight on whether there is a size threshold for anti-ferromagnetic bulk material as CoO to become anti-ferromagnetic. The entire zoo of the 4d and 4f elements can be investigated, which are, due to the large amount of electrons per atom, not computable including electron correlation effects for theory at the moment. These measurements may still give important input to improve the theoretical code in the future. More experimental characterization of substrates and the influence due to the deposited material should be undertaken. In case of small effects or low photo absorption cross sections as for 4d and 4f elements in the soft x-ray regime it may only be possible to measure at more stable storage rings as for example SLS, ESRF and soon PETRA III. The smaller the particle beam position fluctuations of the storage ring, the better for this kind of investigations. A combination of the cluster experiment with a fast switching external magnet ($\approx$ 2 T), would allow to switch the magnetization at each point, which would reduce the requirements on long term stability of the storage ring. Alternatively a very fast switching undulator at a storage ring could provide the same advantage, but is seldom available. Combining the cluster experiments with strong external magnets (7 to 10 T), would allow to measure deposited clusters on non magnetic substrates, opening the field to almost not interacting substrates as oxides, which is bridging the gap towards free cluster measurements. It would be very advantageous, to combine high resolution scanning tunneling microscope (HRSTM) with the cluster experiments. Investigating deposited clusters with HRSTM concerning their arrangement on the surface and their individual moments and to measure the ensemble average with XMCD would be a perfect complementation and would improve comparability to theory a lot. HRSTM investigations a smooth surfaces in needed to distinguish between different cluster sizes [71], while at this point the magnetic substrates used are rather rough, which makes any size inspecting HRSTM measurements most challenging if not even impossible. Measurements of clusters on smooth surfaces as used in HRSTM measurements could be done using of a strong external magnet. This fusion of integrating x-ray spectroscopic investigation techniques with a local probe as HRSTM will be an ambitious aim on the experimental side in the coming years.

Appendix A

Mass spectra

All mass selected clusters measured in this work were created by sputtering a solid target with 30 keV xenon ions. The sputtering process itself is quite complex and since early experiments by Honig in 1958 [63] much investigated, while a good overview can be found in [62]. Typical mass spectra follow an exponential decay with increasing mass [131], while often there are effects of odd even oscillations and some other more stable structures due to electronic shell closure. Above that geometrically highly symmetric structures (magic cluster sizes at geometric shell closures) as for 13 and 55 atoms usually proof very stable and thus stick out by higher sputter yields. Most sputtered clusters are neutral, while charged clusters can easily be extracted by applying a voltage to the target, as it is done in the cluster source (ICARUS) used for this work. All cluster mass spectra were taken for positively charged clusters, of which the dominating portion are singly charged cluster ions.

For the deposition of mass selected Co_nPt_m clusters, two targets with different stoichiometric ratios of Co and Pt were used. The Co/Pt ratio in atomic percent were 25/75 and 50/50 and both had a purity of 99.95%. The mass spectra taken of the two different targets had similar yields for the small clusters, but partially inverting intensity ratios of succeeding sizes of heavier clusters, as can be seen in figure A.3. The stoichiometry of the sputter target affects the sputter yield of the cluster ions. Obviously $Co_1Pt_2^+$ (mass $\approx$ 450 u) and heavier clusters have higher cluster ion currents for the $Co_{25}Pt_{75}$ target. The $Co_{25}Pt_{75}$ target produces decreasing cluster currents from Co_1Pt_x to Co_2Pt_x, while at the $Co_{50}Pt_{50}$ sputter target the current doubles there.

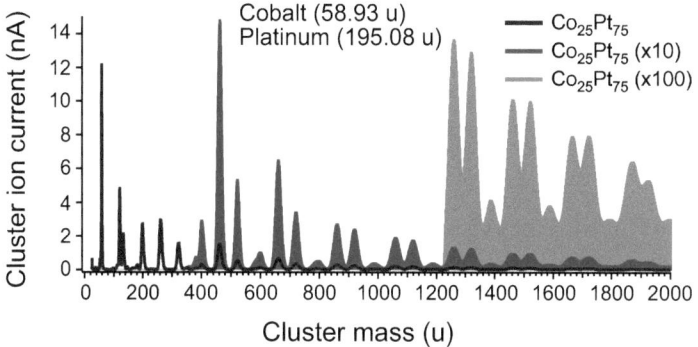

Figure A.1: Mass spectrum of $Co_{25}Pt_{75}$

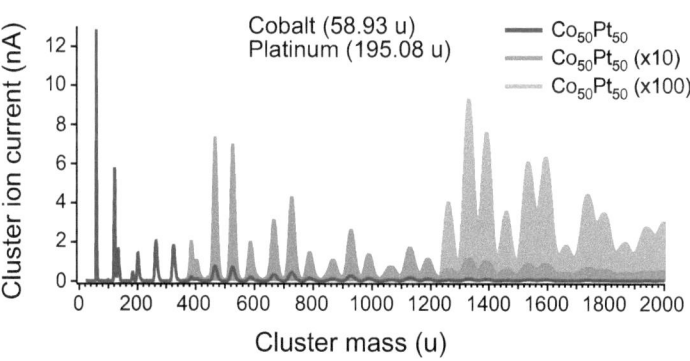

Figure A.2: Mass spectrum of $Co_{50}Pt_{50}$.

Trying to explain the trends with the usual jellium model [1] there would be an electronic shell closure for the lightest clusters $Pt_1Co_1^+$ (18 delocalized electrons), which would favor it from $Pt_1Co_2^+$ (27 delocalized electrons). For the intermediate sizes the jellium model will give neither shell closures for $Pt_nCo_1^+$ nor $Pt_nCo_2^+$ and thus favor nor explain any trend. Even worse is the situation for the heavier clusters in the spectrum, where for the $Co_{50}Pt_{50}$ sputter target the $Pt_nCo_2^+$ is the most common in the experiment, while the jellium model would favor the shell closures at 58 ($Pt_5Co_1^+$) and 68 ($Pt_6Co_1^+$). Just approaching the spectra from a point of stoichiometric distribution of the elements involved one can understand that $Pt_2Co_1^+$ and $Pt_3Co_1^+$ are more favorable in case of the $Co_{25}Pt_{75}$ sputter target and that for larger clusters an additional Co ad-atom is more likely. For the $Co_{50}Pt_{50}$ sputter target the argument of more Co in the sputter fragments would already to be considered at $Pt_3Co_x^+$. Unexplained would remain why $Pt_nCo_1^+$ and $Pt_nCo_2^+$ are the two dominant sputter classes in the spectra, although it appears as if for larger clusters in the case of the $Co_{50}Pt_{50}$ sputter target the peak of Pt_{n+1}^+ is dominated by the $Pt_nCo_3^+$ peak, which is at the same position (within the mass resolution of the system). Such an increase of the Co_3 containing fragments with a decreasing of the Co_1 containing clusters for the sputter target that contains twice as much Co per volume (or three times as much Co per Pt) at least supports the approach of including the stoichiometry of the targets. Co_1^+ and Co_2^+ produce for both sputter targets large cluster currents [2], while the currents for pure Pt clusters are almost negligible, which can not be explained by the weight difference of the sputter gas Xe and the target materials Co and Pt, since $Pt_nCo_1^+$ is even heavier than Pt_x^+, but obviously more in the spectrum. The amount of clusters with more than 3 Co atoms (Pt_nCo_{3+m}) can not be estimated, since they coincide in the mass spectrum with the next larger Pt cluster with 3 Co atoms less ($Pt_{n+1}Co_m$). Nevertheless the first experiments with sputtering alloys with this cluster source was successful and has encouraged further investigations with CoPd and FePt sputter targets so far.

The broader the mass distribution of the natural isotopes an element, the broader

---

1. In the jellium model all lightly bound electrons are assumed to be delocalized and move in a uniform potential field created by the core electrons and the nuclei. In DFT the LDA correlates to this picture.

2. ~15 nA is the highest cluster current so far measured with that sputter source for a pure metal target

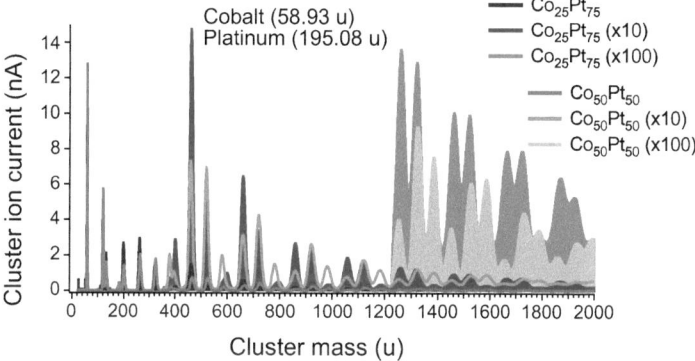

Figure A.3: The stoichiometry of the sputter target influences the yield of the produced clusters. Comparing the mass spectra of $Co_{25}Pt_{75}$ and $Co_{50}Pt_{50}$, it is most obvious that the cluster current heavier clusters is quite different. Co_1Pt_2 (mass ≈ 450) and heavier clusters show much higher currents for the $Co_{25}Pt_{75}$ target. Additionally the $Co_{25}Pt_{75}$ target has a decreasing current from Co_1Pt_x to Co_2Pt_x, while at the $Co_{50}Pt_{50}$ sputter target the current doubles at the same transition.

the peaks in the mass spectrum look for each cluster size of this element. Table A.4 shows a list of all natural isotopes of the elements sputtered for this work. Cobalt belongs to the few mono-isotopic elements and is not very reactive, while Pt being even more inert and rather heavy with only 3 adjacent isotopes contributing to over 90% of the abundance, the CoPt-alloy targets can be considered user friendly sputter material.

During the measurements done for this work several other mass spectra were recorded, since at that point a proof of principle had to be presented, that the experimental setup was able to produce $4d$ and $4f$ transition metal clusters with sufficient cluster yields. The following figures show it has proven possible, although in the case of ytterbium (figure A.12) for instance the reactivity of the sputter material is too high, that the sputter yield may consist to a large extend out of oxidized clusters. Typically as shown at the example of cerium (figure A.8) the oxide layer is only thin if present at all and can easily be removed by sputtering a short while (2-3 minutes in this case).

The best mass resolution of the cluster mass selection is $\frac{m}{\Delta m} \approx 50$. In case of Ni_1 and Cu_1 Δm is about 1.5 u. The mass difference between the lighter Cu isotope ^{63}Cu and the heavier of the two common Ni isotopes ^{60}Ni is 3 u, the mass spectrum could drop down to 1/20 of the Ni peak height before the Cu peak rises, taking the contribution of the remaining heavier isotopes of Ni ($\approx 5\%$) into account. In the mass spectrum of $Cu_{60}Ni_{40}$ displayed in figure A.5 one can very well observe the two peaks of Ni_1 and Cu_1. Obviously the settings used were slightly below the optimum.

isotope	abundance (%)	weight (u)	isotope	abundance (%)	weight (u)
^{54}Fe	5.82	53.939612	^{104}Ru	18.60	103.905424
^{56}Fe	91.18	55.934939	^{136}Ce	0.19	135.907140
^{57}Fe	2.10	56.935396	^{138}Ce	0.25	137.905985
^{58}Fe	0.28	57.933277	^{140}Ce	88.48	139.905433
^{59}Co	100.00	58.933198	^{142}Ce	11.08	141.909241
^{58}Ni	68.27	57.935346	^{141}Pr	100.00	140.907647
^{60}Ni	26.10	59.930788	^{160}Dy	2.34	159.925193
^{61}Ni	1.13	60.931058	^{161}Dy	18.90	160.926930
^{62}Ni	3.59	61.928346	^{162}Dy	25.50	161.926795
^{64}Ni	0.91	63.927968	^{163}Dy	24.90	162.928728
^{63}Cu	69.17	62.939598	^{164}Dy	28.20	163.929171
^{65}Cu	30.83	62.927793	^{168}Yb	0.13	167.933894
^{92}Mo	14.84	91.906808	^{170}Yb	3.05	169.934759
^{94}Mo	9.25	93.905085	^{171}Yb	14.30	170.936323
^{95}Mo	15.92	91.905840	^{172}Yb	21.90	171.936378
^{96}Mo	16.68	91.904678	^{173}Yb	16.12	172.938208
^{97}Mo	9.55	91.906020	^{174}Yb	31.80	173.938859
^{100}Mo	9.63	91.907477	^{176}Yb	12.70	175.942564
^{96}Ru	5.53	95.907599	^{190}Pt	0.01	189.959917
^{98}Ru	1.87	97.905287	^{192}Pt	0.79	191.961019
^{99}Ru	12.70	98.905939	^{194}Pt	32.90	193.962655
^{100}Ru	12.60	99.904219	^{195}Pt	33.80	194.964766
^{101}Ru	17.10	100.905582	^{196}Pt	25.30	195.967315
^{102}Ru	31.60	101.904348	^{198}Pt	7.20	197.967869

Figure A.4: Natural isotopes of the sputter materials and iron

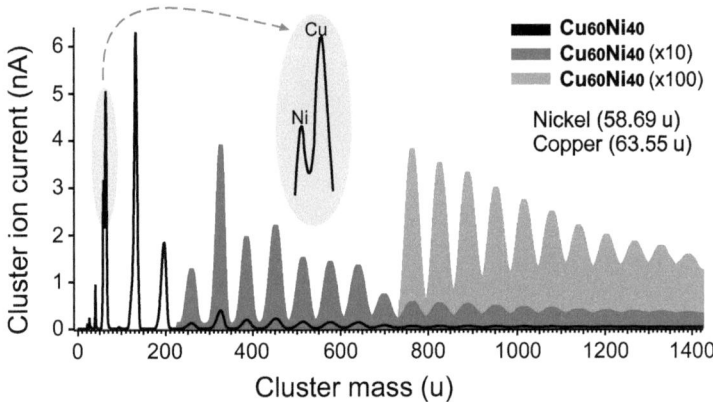

Figure A.5: A Target consisting of $Cu_{60}Ni_{40}$ is a good target to check the resolution of the mass selection. In the blue highlighted ellipse can be seen that Ni_1 and Cu_1 can be separated.

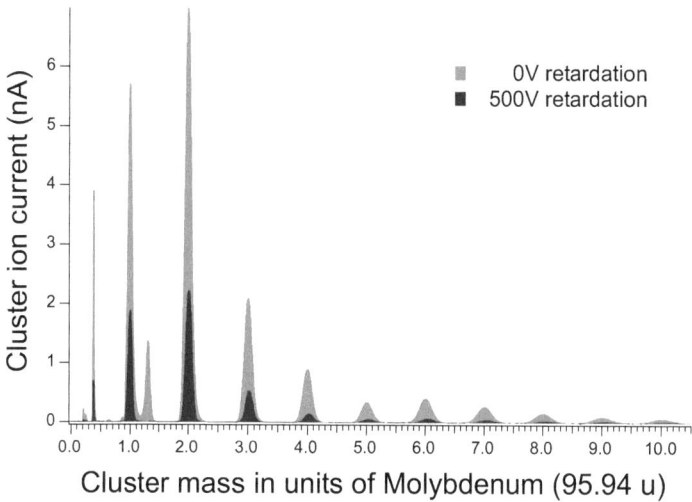

Figure A.6: Mass spectra of 4*d* element molybdenum (Mo), of one is the measured cluster current with the retardation potential (in all measurements in this work: 500 V) applied, the other without the retarding potential. The typical decrease of the cluster current when retarding with the sputter target potential of 500 V is by a factor of 2-3, depending slightly on the sputter target element.

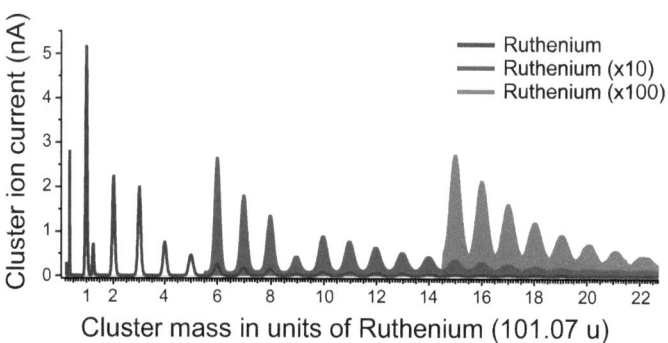

Figure A.7: Mass spectrum of 4*d* element ruthenium (Ru)

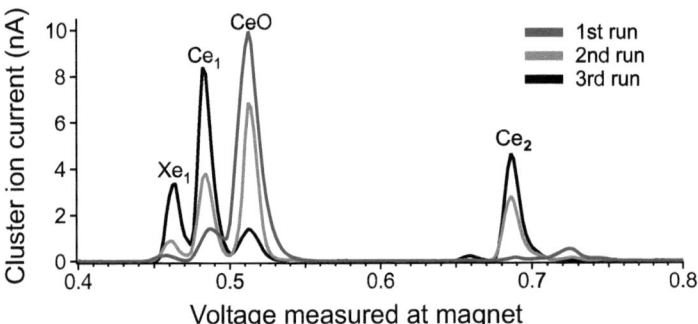

Figure A.8: Time dependent sputter yield: In cases of reactive targets, the surface is always covered with an oxide layer. Sputtering these systems the first mass spectrum has a strongly enhanced proportion of oxidized components. In this example the sputter spot had almost completely eroded the oxide layer after the second short mass spectrum taken, which can be nicely observed by the decreasing intensity of CeO clusters.

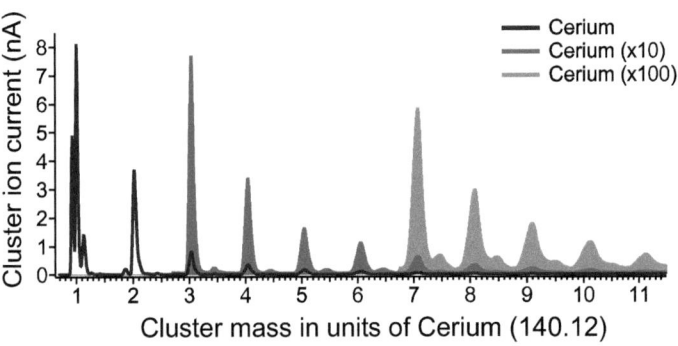

Figure A.9: Mass spectrum of $4f$ element cerium (Ce)

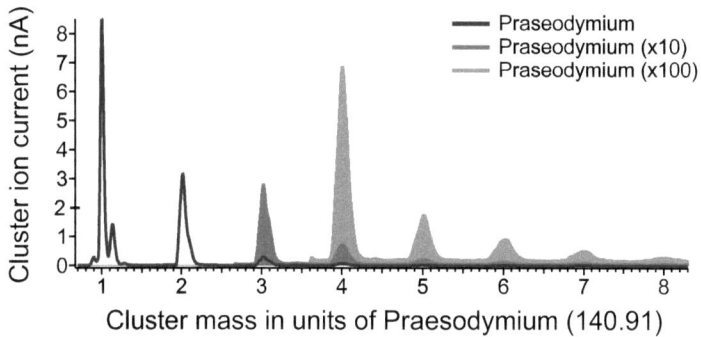

Figure A.10: Mass spectrum of $4f$ element praseodymium (Pr)

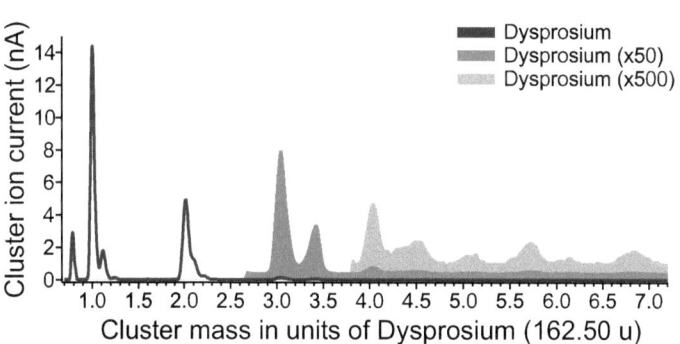

Figure A.11: Mass spectrum of $4f$ element dysprosium (Dy)

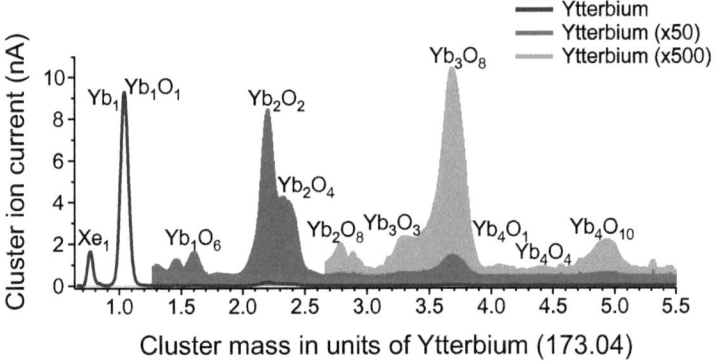

Figure A.12: Mass spectrum of $4f$ element ytterbium (Yb)

Appendix B

Abbreviations

Table B.1: List of all used Abbreviations and what they stand for.

abbreviation	in full
ASA	atomic sphere approximation
ASR	augmented space recursion
BR	branching ratio
CPA	coherent potential approximation
DFT	density functional theory
DMFT	dynamical mean field theory
DOS	density of states
DUV	deep ultra violett (193 nm/6.4 eV)
EUV	Extreme ultra violett (13.5 nm/92 eV)
EXAFS	extended x-ray absorption fine structure
FLAPW	full potential augmented plane wave
FP-LMTO	full potential linearized muffin tin orbital
GGA	generalized gradient approximation
GMR	giant magneto resistance

table ends on page XIV

	table starts on page XII
abbreviation	in full
HF	Hartree Fock
ICARUS	Ionic Clusters by ARgon spUttering Scource
KKR	Korringa-Krohn-Rostocker
LB	Langmuir Blodget
LDA	local density approximation
LEED	low energy electron diffraction
LSDA	local spin density approximation
MAE	magnetic anisotropy energy
MBE	molecular beam epitaxy
MEXAFS	magnetic extended x-ray absorption fine structure
MFA	mean field approximation
MOKE	magneto optical Kerr effect
NEXAFS	near edge x-ray absorption fine structure
OPT	orbital polarization term
PES	photo electron spectroscopy
PMA	perpendicular magnetic anisotropy
SCF	self consistent field
SDW	spin density wave
SEM	scanning electron microscope
SPMDS	spin polarized metastable deexcitation spectroscopy
SRT	spin reorientation transition
STM	scanning tunneling microscope
TB-LMTO	tight binding linearized muffin tin orbital
	table ends on page XIV

table starts on page XII	
abbreviation	in full
TDS	thermal desorption spectroscopy
tey	total electron yield
TSP	titanium sublimation pump
UHV	ultra high vacuum
XAS	x-ray absorption spectroscopy
XMCD	x-ray magnetic circular dichroism
XPS	x-ray photoelectron spectroscopy

List of Figures

1.1	Hard Drive grain sizes	4
1.2	Perpendicular Recording	5
1.3	ITRS Roadmap 2008	7
1.4	Quantum size effect	13
2.1	Scematic XAS spectrum	19
2.2	Relative Auger yield	21
2.3	Electron mean free path	22
2.4	Absorption in single particle picture	24
2.5	XMCD in a simple picture	26
2.6	Absorption of circular light at the L-edges	27
2.7	Thermal desorption spectroscopy	33
2.8	Soft landing of dimers	35
2.9	STM investigation of clusters	36
2.10	Branching ratio of $3d$ transition elements	39
2.11	Effects on spectral energy levels	40
2.12	s-d hybridization of free Co clusters	42
2.13	Theoretical predictions for CoPt alloys	44
2.14	$L1_0$-structure and $L1_2$-structure	46
2.15	Free and deposited Co	46
2.16	Thin film magnetization of Ni on Cu(100)	50
2.17	Co capping layer on Fe thin film	52
2.18	Thin film magnetization of Fe on Cu(100)	53
2.19	Depth resolved magnetization of Fe on Cu(100)	54
3.1	Structure of a CoPt nanoparticles	59
3.2	Temperature dependent synthesis	60

3.3	Nucleation process in CoPt nanoparticles Synthesis	61
3.4	Structural formula of essential chemicals	61
3.5	Post synthesis process	62
3.6	Particle deformation upon deposition	63
3.7	Magnetic saturation of wet chemical nanoparticles	64
3.8	ICARUS Sputter source	66
3.9	CoPt mass spectrum	68
3.10	Spacial distribution of deposited clusters	69
3.11	ICARUS spectroscopy section	72
3.12	Manipulator head of ICARUS	74
3.13	Spectroscopy chamber of ICARUS	75
3.14	Mass selected cluster preparation cycle	76
3.15	Beamline UE56-PGM	80
3.16	Measuring chamber at synchrotron	81
3.17	Different measuring geometries	82
3.18	Sketch of measuring procedure	83
3.19	Specifications of EMP/2 x-file	84
3.20	Normalization and background treatment	86
3.21	Mass selected cluster background	88
3.22	Raw data of deposited mass selected clusters	89
4.1	Sample storage N_2 vs air	92
4.2	Defining oxidation measure Co/CoO	94
4.3	Oxidation reference curve	95
4.4	Oxidation of the nanoparticles with age	96
4.5	Oxidation of the nanoparticles with size	97
4.6	Oxidation of 10 to 15 days old sample with age and size	99
4.7	Simple particle oxidation model	100
4.8	Simple model calculations	101
4.9	Enhanced oxidation models	101
4.10	Fixed core particle model	105
4.11	Flat layer and macro crystals	106
4.12	Flat layer preparation and macro crystals	108
4.13	SO-splitting and BR	109
4.14	Oxidation effects ml to ms	111

4.15	Angular dependence of ratio ml to ms	112
4.16	Overview measurements vs literature	113
4.17	Spin and orbital moment versus age	114
4.18	Spin and orbital moment versus age	115
4.19	Changes of orbital moment with angle of incidence	116
4.20	Angular dependence of ml and ms for old particles	117
5.1	NEXAFS and magnetic NEXAFS of Fe on Cu(111)	130
5.2	Fe background evolution due to Co coverage	132
5.3	Fe EXAFS background for Co	134
5.4	Background treatment for cluster on an Fe substrate - part I	135
5.5	Step function removal and the cutoff alternative	136
5.6	Background treatment for cluster on an Fe substrate - part II	137
5.7	Fit masks for Fe on Cu(100) background treatment	137
5.8	Fe film coverage estimation	139
5.9	Fe film thickness on Cu(100)	140
5.10	Oxidation of mass selected particles	142
5.11	Branching ratio of L_3 to L_2 edge	144
5.12	Apparent $2p$ spin orbit splitting	145
5.13	Dichroism of Co monomer, dimer and trimer on Fe/Cu(100)	147
5.14	Dichroism of Co monomer and dimer on Ni/Cu(100)	150
5.15	Dichroism of Co monomer, while varying Pt	151
5.16	Dichroism of Co on Ni/Cu(100)	153
5.17	Co_1 on Fe, Ni and Pt	156
A.1	Mass spectra $Co_{25}Pt_{75}$ atomic percent	II
A.2	Mass spectra $Co_{50}Pt_{50}$ atomic percent	II
A.3	Comparing different CoPt mass spectra	IV
A.4	Natural isotopes of the sputter materials	VI
A.5	Mass spectrum Cu/Ni alloy	VII
A.6	Mass spectrum molybdenum (Mo)	VIII
A.7	Mass spectrum ruthenium (Ru)	VIII
A.8	Mass spectra time dependency	IX
A.9	Mass spectrum cerium (Ce)	IX
A.10	Mass spectrum praseodymium (Pr)	X

A.11 Mass spectrum dysprosium (Dy) X
A.12 Mass spectrum ytterbium (Yb) XI

List of Tables

2.1	Properties of Fe, Co, Ni and Cu	17
2.2	3d metal interaction	37
3.1	Beamline specifications	79
4.1	Fresh nanoparticle oxidation state	102
4.2	Aged nanoparticle oxidation state	104
4.3	Magnetic moments of Co_xPt_{100-x} nanoparticles	120
4.4	Self absorption correction	121
4.5	Magnetic moments of Co_xPt_{100-x} nanoparticles tey corrected	122
4.6	Magnetic moments oxidized wet chemicals tey corrected	124
4.7	Magnetic moments of experimental literature references I	125
4.8	Magnetic moments of theory literature references I	126
5.1	Magnetic moments of Fe/Cu(100)	133
5.2	Magnetic moments of clusters on Fe/Cu(100)	154
5.3	Magnetic moments of clusters on Ni/Cu(100)	154
5.4	Magnetic moments of oxidized clusters on nickel	154
5.5	Magnetic moments of experimental literature references II	159
5.6	Magnetic moments of theory literature references II	160
B.1	Abbreviations	XII

Bibliography

[1] F. Aguilera-Granja, A. Garcia-Fuente, and A. Vega. Comparative ab initio study of the structural, electronic, and magnetic trends of isoelectronic late 3d and 4d transition metal clusters. *Physical Review B (Condensed Matter and Materials Physics)*, 78(13):134425, 2008.

[2] V. T. Aleksandrovic. *$CoPt_3$ Nanoparticles: Ligand Exchange and Film Preparation*. PhD thesis, University of Hamburg, 2006.

[3] M. Altarelli. Orbital-magnetization sum rule for x-ray circular dichroism: A simple proof. *Phys. Rev. B*, 47:597–598, 1993.

[4] M. Altarelli. Sum rules x-ray magnetic circular dichroism. *Il Nuovo Cimento*, 20:1067–1073, 1998.

[5] K. Amemiya, S. Kitagawa, D. Matsumura, H. Abe, T. Ohta, and T. Yokoyama. observation of magnetic depth profiles of thin fe films on cu(100) and ni/cu(100) with the depth-resolved x-ray magnetic circular dichroism. *Appl. Physics Lett.*, 84:936–938, 2004.

[6] K. Amemiya, S. Kitagawa, D. Matsumura, T. Yokoyama, and T. Ohta. Development of a depth-resolved x-ray magnetic circular dichroism: application to fe/cu(100) ultrathin films. *JoPhys.:Cond.Matt.*, 15:S561–S571, 2003.

[7] K. Amemiya, T. Yokoyama, Y. Yonamoto, D. Matsumura, and T. Ohta. O k-edge x-ray magnetic circular dichroism of atomic o adsorbed on an ultrathin co/cu(100) film: Comparison with molecular co on co/cu(100). *Phys. Rev. B*, 64:1324051–1324054, 2001.

[8] C. Andersson, B. Sanyal, O. Eriksson, L. Nordström, O. Karis, D. Arvanitis, T. Konishi, E. Holub-Krappe, and J. Hunter Dunn. Influence of ligand states on the relationship between orbital moment and magnetocrystalline anisotropy. *PRL*, 99:177207, 2007.

[9] A.L. Ankudinov, A.I. Nesvizhskii, and J.J. Rhehr. Dynamic screening effects in x-ray absorption spectra. *Phys. Rev. B*, 67:1151201–1151206, 2003.

[10] K. Baberschke. Ferromagnetic monolayers: a fresh look at fundamentals. *phys. stat. sol.*, 236:233–239, 2003.

[11] S.D. Bader. Magnetism in low dimensionality. *Surface Science*, 500:172–188, 2002.

[12] J. Bansmann, M. Getzlaff, A. Kleibert, F. Bulut, R.K. Geberhardt, and K.H. Meiwes-Broer. Mass-filtered cobalt clusters in contact with epitaxially ordered metal surfaces. *Appl.Phys. A*, 82:73–79, 2006.

[13] J. Bansmann, A. Kleibert, F. Bulut, M. Getzlaff, P. Imperia, C. Boeglin, and K.-H. Meiwes-Broer. Temperature dependent magnetic spin and orbital moments of mass-filtered cobalt clusters on au(111). *Eur.Phys.J.D.*, 45:521–528, 2007.

[14] D. Bazin, I. Kovács, L. Guczi, P. Parent, C. Laffon, F. De Groot, O. Ducreux, and J. Lynch. Genesis of co/sio$_2$ catalysts: Xas study at the cobalt l$_{III,II}$ absorption edges. *Journal of Catalysis*, 189:456–462, 2000.

[15] Albert Biedermann, Rupert Tscheließnig, Michael Schmid, and Peter Varga. Crystallographic structure of ultrathin fe films on cu(100). *Phys. Rev. Lett.*, 87(8):086103, Aug 2001.

[16] I.M.L. Billas, A. Châtelain, and W.A. de Heer. Magnetism from the atom to the bulk in iron, cobalt, and nickel. *Science*, 265:1682–1684, 1994.

[17] I.M.L. Billas, A. Châtelain, and W.A. de Heer. Magnetism of fe, co and ni clusters in molecular beams. *Journal of Magnetism and Magnetic Materials*, 168:64–84, 1997.

[18] G. Binasch, P. Grünberg, F. Saurenbach, and W. Zinn. Enhanced magnetoresistance in layered magnetic structures with antiferromagnetic interlayer exchange. *Phys. Rev. B*, 39(7):4828–4830, Mar 1989.

[19] S. Bornemann. private communication.

[20] K. Bromann, H. Brune, C. Féix, W. Harbich, R. Monot, J. Buttet, and K. Kern. Hard and soft landing of mass selected ag clusters on pt(111). *Surf.Sci.*, 377-379:1051–1055, 1997.

[21] K. Bromann, C. Féix, H. Brune, W. Harbich, R. Monot, J. Buttet, and K. Kern. Controlled deposition of size-selected silver nanoclusters. *Science*, 274:956–958, 1996.

[22] P. Bruno and J.-P. Renard. Magnetic surface anisotropy of transition metal ultrathin films. *Appl. Phys. A*, 49:499–506, 1989.

[23] J.P. Bucher, D.C. Douglass, and L.A. Bloomfield. Magnetic properties of free cobalt clusters. *Phys. Rev. Lett.*, 66(23):3052–3055, 1991.

[24] G. S. Chang, Y. P. Lee, J. Y. Rhee, J. Lee, K. Jeong, and C. N. Whang. Realization of a large magnetic moment in the ferromagnetic copt bulk phase. *Phys. Rev. Lett.*, 87(6):067208, Jul 2001.

[25] C.T. Chen, Y.U. Idzerda, H.-J. Lin, N.V. Smith, G. Meigs, E. Chaban, G.H. Ho, E. Pellegrin, and F. Sette. Experimental confirmation of x-ray magnetic circular dichroism sum rules for iron and cobalt. *Phys. Rev. Lett.*, 75(1):152–155, 1995.

[26] H.-P. Cheng and U. Landman. Controlled deposition, soft landing, and glass formation in nanocluster-surface collisions. *Science*, 260:1304–1307, 1993.

[27] H.-P. Cheng and U. Landman. Controlled deposition and glassification of copper nanoclusters. *J.Phys.Chem.*, 98:3527–3537, 1994.

[28] F. M. F. de Groot, J. C. Fuggle, B. T. Thole, and G. A. Sawatzky. 2p x-ray absorption of 3d transition-metal compounds: An atomic multiplet description including the crystal field. *Phys. Rev. B*, 42(9):5459–5468, Sep 1990.

[29] J. Hunter Dunn, D. Arvanitis, R. Carr, and N. Mårtensson. Magnetisation reorientation in ultra-thin fe films on cu(100) upon deposition of co. *J. Syn. Rad.*, 8:463–465, 2001.

[30] H. A. Dürr, S. S. Dhesi, E. Dudzik, D. Knabben, G. van der Laan, J. B. Goedkoop, and F. U. Hillebrecht. Spin and orbital magnetization in self-assembled co clusters on au(111). *Phys. Rev. B*, 59(2):R701–R704, Jan 1999.

[31] H.A. Dürr, G. van der Laan, J. Vogel, G. Panaccione, N.B. Brookes, E. Dudzik, and R. McGrath. Enhanced orbital magnetism at the nanostructured co/cu(1 1 13) surface. *Phys. Rev. B*, 58:R11853–R11856, 1998.

[32] D.A. Eastham and I.W. Kirkman. Highly enhanced orbital magnetism on cobalt cluster surfaces. *J.Phys.:Condens.Matter*, 12:L525–L532, 2000.

[33] H. Ebert. *Circular Magnetic X-ray Dichroism in Transition Metal Systems, volume 466 of lecture notes in Physics*. Springer, Berlin Heidelberg, 1996.

[34] C. Ederer, M. Komelj, J. W. Davenport, and Fähnle. Comment on the analysis of angle-dependent x-ray magnetic circular dichroism in systems with reduced dimensionality. *J. Electron Spectr. and Rel. Phenom.*, 130:97–100, 2003.

[35] C. Ederer, M. Komelj, and M. Fähnle. Magnetism in systems with various dimensionalities: A comparison between fe and co. *Phys.Rev. B*, 68:524021–524024, 2003.

[36] Claude Ederer, Matej Komelj, Manfred Fähnle, and Gisela Schütz. Theory of induced magnetic moments and x-ray magnetic circular dichroism in co-pt multilayers. *Phys. Rev. B*, 66(9):094413, Sep 2002.

[37] H.-U. Ehrke. *Mononumerische Metallcluster: Erzeugung und Röntgenspektroskopische Messungen*. PhD thesis, Technische Universität München, 2000.

[38] Olle Eriksson, Börje Johansson, R. C. Albers, A. M. Boring, and M. S. S. Brooks. Orbital magnetism in fe, co, and ni. *Phys. Rev. B*, 42(4):2707–2710, Aug 1990.

[39] U. Fano. Spin orientation of photoelectrons ejected by circular polarized light. *Phys. Rev.*, 178:131–178, 1969.

[40] U. Fano. Spin orientation of photoelectrons: Erratum and addendum. *Phys.Rev.*, 184:250, 1969.

[41] K. Fauth. How well does total electron yield measure x-ray absorption in nanoparticles? *Appl.Phys. Lett.*, 85(15):3271–3273, 2004.

[42] L. Favre, V. Dupuis, E. Bernstein, P. Mélinon, and Pérez. Structural and magnetic properties of copt mixed clusters. *Phys.Rev. B*, 74:144391–144399, 2006.

[43] S. Fedrigo, W. Harbich, and J. Buttet. Soft landing and fragmentation of small clusters deposited in noble-gas films. *Phys. Rev. B*, 58(11):7428–7433, Sep 1998.

[44] Rui-Juan Feng, Xiao-Hong Xu, and Hai-Shun Wu. Electronic structure and magnetism in (copt)n(n[less-than-or-equals, slant]5) clusters. *Journal of Magnetism and Magnetic Materials*, 308(1):131 – 136, 2007.

[45] A. Fert, P. Grünberg, A. Barthélémy, F. Petroff, and W. Zinn. Layered magnetic structures: interlayer exchange coupling and giant magnetoresistance. *Journal of Magnetism and Magnetic Materials*, 140-144(Part 1):1 – 8, 1995. International Conference on Magnetism.

[46] P.Le Fevre, H. Magnan, and D. Chandesris. Tetragonal structure of thin nickel films on cu(100). *The European Physical Journal B*, 10:555–562, 1999.

[47] J. Fink, Th. Müller-Heinzerling, and B. Scheerer. 2p absorption spectra of the 3d elements. *Phys. Rev. B*, 32:4899–4904, 1985.

[48] D.E. Fowler and J.V. Barth. Magnetic anisotropy of glide-distorted fcc and of bcc ultrathin fe/cu(001) films. *Phys. Rev. B*, 53(9):5563–5569, 1996.

[49] I. Galanakis, M. Alouani, and H. Dreysse. Calculated magnetic properties of low-dimensional systems: the aucu- and aucu$_3$-type ferromagnets. *Journal of Magnetism and Magnetic Materials*, 242-245:27–32, 2002.

[50] P. Gambardella, M. Russo, M. Veronese, S.S. Dhesi, C. Grazioli, A. Dallmeyer, I. Cabria, R. Zeller, P.H. Dederichs, K. Kern, C. Carbone, and H. Brune. Giant magnetic anisotropy of single atoms and nanoparticles. *Science*, 300:1130, 2003.

[51] M. Getzlaff, J. Bannsmann, F. Bulut, R.K. Gebhardt, A. Kleibert, and K.H. Meiwes-Broer. Structure, composition and magnetic properties of size-selected feco alloy clusters on surfaces. *Appl.Phys. A*, 2006.

[52] W. Grange, I. Galanakis, M. Alouani, M. Maret, J.-P. Kappler, and A. Rogalev. Experimental and theoretical x-ray magnetic-circular-dichroism study of the magnetic properties of $co_{50}pt_{50}$ thin films. *Phys. Rev. B*, 62(2):1157–1166, 2000.

[53] W. Grange, M. Maret, J.-P Kappler, J. Vogel, A. Fontaine, F. Petroff, G. Krill, A. Rogalev, J. Goulon, M. Finazzi, and N. B. Brookes. Magnetocrystalline anisotropy in (111) $copt_3$ thin films probed by x-ray magnetic circular dichroism. *Phys. Rev. B*, 58(10):6298–6304, Sep 1998.

[54] G. Y. Guo, H. Ebert, W. M. Temmerman, and P. J. Durham. First-principles calculation of magnetic x-ray dichroism in fe and co multilayers. *Phys. Rev. B*, 50(6):3861–3868, Aug 1994.

[55] C. Gustin, L.H.A. Leunissen, A. Mercha, S. Decoutere, and G. Lorusso. Impact of line width roughness on the matching performances of next-generation devices. *Thin solid Films*, 516:3690–3696, 2007.

[56] K. Heinz, S. Müller, and L. Hammer. Crystollography of ultrathin iron, cobalt and nickel films grown epitaxially on copper. *J. Phys.: Condens. Matter*, 11:9437–9454, 1999.

[57] B. Hernnäs, M. Karolewski, H. Tillborg, A. Nilsson, and N. Mårtensson. On the growth of ni on cu(100). *Surface Science*, 302:64–72, 1994.

[58] F. Hippert, E. Geissler, J.L. Hodeau, E. Lelièvre-Berna, and J.-R. Regnard. *Neutron and X-ray Spectoscopy*. Springer Verlag, 2006.

[59] Hitachi. http://www.hitachigst.com.

[60] O. Hjortstam, K. Baberschke, J.M. Wills, B. Johansson, and O. Eriksson. Magnetic anisotropy and magnetostriction in tetragonal and cubic ni. *Phys. Rev. B*, 55:15026–15032, 1997.

[61] O. Hjortstam, J. Trygg, J.M. Wills, B. Johansson, and O. Eriksson. Calculated spin and orbital moments in the surfaces of the 3d metals fe, co, ni and their overlayers on cu(100). *Phys. Rev. B*, 53(14):92049213, 1996.

[62] W.O. Hofer. *Sputtering by Particle Bombardement III*, chapter 2.Angular, Energy and Mass Distribution of Sputtered Particles, pages 15–90. Springer Verlag, 1991.

[63] R.E. Honig. Sputtering of surfaces by positive ion beams of low energy. *Journal of Applied Physics*, 29:549–554, 1958.

[64] Horngming Hsieh, R. S. Averback, Harrell Sellers, and C. P. Flynn. Molecular-dynamics simulations of collisions between energetic clusters of atoms and metal substrates. *Phys. Rev. B*, 45(8):4417–4430, Feb 1992.

[65] http://optics.org. Ibm beats optical lithography limits. http://optics.org, Februar 2006.

[66] J. Hunter, D. Arvanitis, and N. Mårtensson. Magnetism of thin fe films on cu(100). *Phys. Rev. B*, 54(16):R11157–R11160, 1996.

[67] J. Hunter-Dunn, D. Arvanitis, N. Mårtensson, M. Tischer, F. May, M. Russo, and K. Baberschke. An angle-dependent magnetic circular x-ray dichroism study of co/cu(100): experiment versus theory. *J.Phys.: Condens. Matter*, 7:1111–1119, 1995.

[68] S. Imada, T. Shishidou, T. Muro, S. Suga, H. Maruyama, K. Kobayashi, H. Yamazaki, and T. Kanomata. Mcd in co and pt soft x-ray absorption in $copt_3$. *Physica B*, 237-238:369–371, 1997.

[69] P. Imperia. private communication.

[70] P. Imperia, L. Glaser, M. Martins, P. Andreazza, J. Penuelas, V. Alesandrovic, H. Weller, C. Andreazza-Vignolle, and W. Wurth. Xmcd studies of $co_x pt_{100-x}$ nanoparticles prepared by vapour deposition and chemical synthesis. *phys. stat. sol. a*, 5:1047–1051, 2008.

[71] ITRS. 2008 litho itrs meeting. http://www.itrs.net, 7 2008.

[72] V. Jähnke, U. Conrad, J. Güdde, and E. Matthias. Shg investigations of the magnetization of thin ni and co films on cu(001). *Applied Physics B*, 68:485–489, 1999.

[73] A. Kleibert, J. Passig, K.-H. Meiwes-Broer, M. Getzlaff, and J. Bansmann. Structure and magnetic moments of mass-filtered deposited nanoparticles. *Journal of Applied Physics*, 101(11):114318, 2007.

[74] J. Kliewert, R. Berndt, J. Minar, and H. Ebert. Scanning tunnelling microscopy and electronic structure of mn clusters on ag(111). *Appl. Phys. A*, 82:63–66, 2006.

[75] M. Komelj, C. Ederer, J. W. Davenport, and Fähnle. From the bulk to monoatomic wires: An ab initio study of magnetism in co systems with various dimensionality. *Phys.Rev. B*, 66:1404071–1404074, 2002.

[76] M. Kurahashi, T. Suzuki, X. Ju, and Y. Yamauchi. Spin-polarized metastable-atom deexcitation spectroscopy of fe/cu(100) surfaces with perpendicular magnetization. *Phys. Rev. B*, 67(2):024407, Jan 2003.

[77] J.T. Lau. *Magnetische Eigenschaften kleiner massenselektierter Übergangsmetallcluster*. Desy-Thesis-2002-016, Notkestr. 85, 22607 Hamburg, 2002.

[78] J.T. Lau, A. Achleitner, H.-U. Ehrke, U. Langenbuch, M. Reif, and W. Wurth. Ultrahigh vacuum cluster deposition source for spectroscopy with synchrotron radiation. *Rewiev of Scientific Instruments*, 76:063902–1, 2005.

[79] J.T. Lau, A. Achleitner, H.-U. Ehrke, and W. Wurth. Soft landing of size-selected clusters in rare gas matrices. *Low Temperature Physics*, 29:223–227, 2003.

[80] J.T. Lau, A. Föhlisch, M. Martins, R. Nietubyc, M. Reif, and W. Wurth. Spin and orbital magnetic moments of deposited small iron clusters studied by x-ray magnetic circular dichroism spectroscopy. *New Journal of Physics*, 4:98.1, 2002.

[81] J.T. Lau, A. Föhlisch, R. Nietubyc, M. Reif, and W. Wurth. Size-dependant magnetism of deposited small iron clusters studied by x-ray magnetic circular dicroism. *Phys. Rev. Lett.*, 89(5):57201, 2002.

[82] J.T. Lau, J. Rittmann, V. Zamudio-Bayer, M. Vogel, K. Hirsch, Ph. Klar, F. Lofink, and T. Möller. Size dependence of $l_{2,3}$ branching ratio and 2p core-hole screening in x-ray absorption of metal clusters. *Phys. Rev. Lett.*, 101:1534011–1534014, 2008.

[83] A. Lichtenstein.

[84] S.H. Liou, S. Huang, E. Klimek, and R.D. Kirby. Enhancement of coercivity in nanometer-size copt crystallites. *J. of Appl. Physics*, 85:4334–4336, 1999.

[85] Shu-Rong Liu, Hua-Jin Zhai, and Lai-Sheng Wang. Electronic and structural evolution of co_n clusters ($n = 1 − 108$) by photoelectron spectroscopy. *Phys. Rev. B*, 64(15):153402, Sep 2001.

[86] Shu-Rong Liu, Hua-Jin Zhai, and Lai-Sheng Wang. s-d hybridization and evolution of the electronic and magnetic properties in small co and ni clusters. *Phys. Rev. B*, 65(11):113401, Feb 2002.

[87] W.L. Liu, K. Alim, A.A. Balandin, D.M. Mathews, and J.A. Dodds. Assembly and characterization of hybrid virus-inorganic nanotubes. *Appl.Phys.Lett.*, 86:253108, 2005.

[88] A. Lobo, H. Borchert, E.V. Shevchenko, R. Doehrmann, S. Adam, H. Weller, and T. Möller. Soft x-ray photoelectron spectroscopy investigations of copt$_3$ nanocrystals. Technical report, HASYLAB, 2003.

[89] S. Lounis, M. Reif, P. Mavropoulos, L. Glaser, P.H. Dederichs, M. Martins, S. Blügel, and W. Wurth. Non-collinear magnetism of cr nanostructures on fe$_3$ml/cu(100): First principles and experimental investigations. *EPL*, 81:47004, 2008.

[90] N. Mårtensson and A. Nilsson. On the origin of core-level binding energy shifts. *J. of electr. Spectr.*, 75:209–223, 1995.

[91] C. Mao, D.J. Solis, B.D. Reiss, S.T. Kottmann, R.Y. Sweeney, A. Hayhurst, G. Georgiou, B. Iverson, and A.M. Belcher. Virus-based toolkit for the directed synthesis of magnetic and semiconducting nanowires. *Science*, 303:213–217, 2004.

[92] M. Marangolo, P. Ohresser, N.B. Brooks, S. Cherifi, C. Boeglin, M. Eddrief, and V.H. Etgens. Tails of near-edge x-ray absorption spectra as fingerprint of magnetic and structural phase transitions. application to metallic 3d ultra thin film. *J. of Appl. Physics*, 93:5151–5155, 2003.

[93] G. Margaritondo. A primer in synchrotron radiation: Everything you wanted to know about sex (synchrotron emission of x-rays) but were afraid to ask. *J.Synchrotron Rad.*, 2:148–154, 1995.

[94] G. Margaritondo. Synchrotron light in a nutshell. *Surf.Rev. and Lett.*, 7:379–387, 2000.

[95] B.I. Min, T. Oguchi, and A.J. Freeman. Structural, electronic, and magnetic properties of co: Evedence for magnetism-stabilizing structure. *Phys. Rev. B*, 33:7852–7854, 1986.

[96] S. Müller, P. Bayer, C. Reischl, K. Heinz, B. Feldmann, H. Zillgen, and M. M. Wuttig. Structural instability of ferromagnetic fcc fe films on cu(100). *Phys. Rev. Lett.*, 74:765–768, 1995.

[97] T. Nakagawa, H. Wantanabe, and T. Yokoyama. Moke and xmcd study on k adsorption on fe ultrathin films on cu(001). *Annual Review 2005*, pages 59–62, 2005.

[98] N. Nakajima, T. Koide, T. Shidara, H. Miyauchi, H. Fukutani, A. Fujimori, K. Iio, T. Katayama, M. Nývlt, and Y. Suzuki. Perpendicular magnetic anisotropy caused by interfacial hybridization via enhanced orbital moment in co/pt multilayers: Magnetic circular x-ray dichroism study. *Phys. Rev. Lett.*, 81(23):5229–5232, Dec 1998.

[99] Reiko Nakajima, J. Stöhr, and Y. U. Idzerda. Electron-yield saturation effects in l-edge x-ray magnetic circular dichroism spectra of fe, co, and ni. *Phys. Rev. B*, 59(9):6421–6429, Mar 1999.

[100] R. Nietubyc, A. Föhlisch, L. Glaser, J.T. Lau, M. Martins, M. Reif, and W. Wurth. L-edge x-ray absorption fine structure study of growth and morphology of ultrathin nickel films deposited on copper. *Phys. Rev. B*, 70:2354141–2354148, 2004.

[101] W.L. O'Brien, T. Droubay, and B.P. Tonner. Transitions in the direction of magnetism in ni/cu(100) ultrathin films and the effects of capping layers. *Phys. Rev. B*, 54:9297–9303, 1996.

[102] W.L. O'Brien, B.P. Tonner, G.R. Harp, and S.S.P. Parkin. Experimental investigation of dichroism sum rules for v, cr, mn, fe, co, and ni: Influence of diffuse magnetism. *J. Appl. Physics*, 76:6462–6464, 1994.

[103] K. Okada and A. Kotani. Complementary roles of co 2p x-ray absorption and photoemission spectra in coo. *J. of the Phys. Soc. of Japan*, 61:449–453, 1992.

[104] D. P. Pappas, C. R. Brundle, and H. Hopster. Reduction of macroscopic moment in ultrathin fe films as the magnetic orientation changes. *Phys. Rev. B*, 45(14):8169–8172, Apr 1992.

[105] D. P. Pappas, K.-P. Kämper, B. P. Miller, H. Hopster, D. E. Fowler, C. R. Brundle, A. C. Luntz, and Z.-X. Shen. Spin-dependent electron attenuation by transmission through thin ferromagnetic films. *Phys. Rev. Lett.*, 66(4):504–507, Jan 1991.

[106] D. Paudayal, T. Saha-Dasgupta, and A. Mookerjee. Magnetic properties of x-pt (x= fe, co, ni) alloy systems. *J.Phys.: Condens. Matter*, 16:2317–2334, 2004.

[107] H. Petersen, M. Willmann, F. Schäfers, and W. Gudat. Circularly polarized soft x-rays in the photon energy range 30-2000 ev at the bessy sx700/3. *Nuclear Instruments and Methods in Physics Research Section A: Accelerators, Spectrometers, Detectors and Associated Equipment*, 333(2-3):594 – 598, 1993.

[108] J.P. Pierce, J. Shen, and Ruqian Wu. Spin reorientation of ultrathin fe/cu(100) films driven by the redistribution of co adatoms. *Phys. Rev. B*, 65:1324081–1324084, 2002.

[109] W. Platow, U. Bovensiepen, P. Poulopoulos, M. Farle, K. Baberschke, L. Hammer, S. Walter, S. Müller, and K. Heinz. Structure of ultrathin ni/cu(001) films as a function of film thickness, temperature, and magnetic order. *Phys. Rev. B*, 59:12641–12646, 1999.

[110] T.J. Regan, H. Ohldag, C. Stamm, F. Nolting, J. Lüning, J. Stöhr, and R.L. White. Chemical effects at metal/oxide interfaces studied by x-ray-absorption spectroscopy. *Phys. Rev. B*, 64:214422, 2001.

[111] M. Reif. *Magnetic properties of small chromium and gadolinium clusters, deposited on magnetic substrates*. PhD thesis, Universität Hamburg, 2005.

[112] R.A. Ristau. *Mcrostructural and Magnetic Characterization of CoPt and FePt thin Films*. PhD thesis, Lehigh University, 1998.

[113] C. O. Rodriguez, M. V. Ganduglia-Pirovano, E. L. Peltzer y Blancá, M. Petersen, and P. Novák. Orbital and dipolar contributions to the hyperfine fields in bulk bcc fe, hcp co, and at the fe/ag(100) interface: The inclusion of orbital polarization. *Phys. Rev. B*, 63(18):184413, Apr 2001.

[114] D. Sander. The correlation between mechanical stress and magnetic anisotropy in ultrathin films. *Rep. Prog. Phys.*, 62:809–858, 1999.

[115] K. J. S. Sawhney, F. Senf, M. Scheer, F. Schäfers, J. Bahrdt, A. Gaupp, and W. Gudat. A novel undulator-based pgm beamline for circularly polarised synchrotron radiation at bessy ii. *Nuclear Instruments and Methods in Physics Research Section A: Accelerators, Spectrometers, Detectors and Associated Equipment*, 390(3):395 – 402, 1997.

[116] A. Scherz, E.K.U. Gross, H Appel, C. Sorg, K. Baberschke, and H. Wende. the kernel of time-dependent density functional theory with x-ray absorption spectroscopy of 3d transition metals. *Phys. Rev. Lett.*, 95:2530061–2530064, 2005.

[117] D. Schmitz, C. Charton, A. Scholl, C. Carbone, and W. Eberhardt. Magnetic moments of fcc fe overlayers on cu(100) and co(100). *Phys. Rev. B*, 59(6):4327–4333, 1999.

[118] G. Schütz, W. Wagner, W. Wilhelm, P. Kienle, R. Zeller, R. Frahm, and G. Materlik. Absorption of circularly polarized x rays in iron. *Phys. Rev. Lett.*, 58(7):737–740, 1987.

[119] J. Schwitalla and H. Ebert. Core-hole interaction in the x-ray absorption spectroscopy of 3d transition metals. *Phys. Rev. Lett.*, 80:4586–4589, 1998.

[120] Sematech. Sematech euv resist at 22nm half-pitch. http://www.semiconductor.net, December 2008.

[121] A. L. Shapiro, P. W. Rooney, M. Q. Tran, F. Hellman, K. M. Ring, K. L. Kavanagh, B. Rellinghaus, and D. Weller. Growth-induced magnetic anisotropy and clustering in vapor-deposited co-pt alloy films. *Phys. Rev. B*, 60(18):12826–12836, Nov 1999.

[122] J. Shen, A.K. Swan, and J.F. Wendelken. Determination of critical thickness of spin reorientation in metastable magnetic ultrathin films. *Appl. Physics Lett.*, 75:2987–2989, 1999.

[123] E. V. Shevchenko, D. V. Talapin, H. Schnablegger, A. Kornowski, Ö. Festin, P. Svedlindh, M. Haase, and H. Weller. Study of nucleation and growth in the organometallic synthesis of magnetic alloy nanocrystals: The role of nucleation rate in size control of copt$_3$ nanocrystals. *Journal of the American Chemical Society*, 125:9090, 2003.

[124] E.V. Shevchenko, D.V. Talapin, A.L. Rogach, A. Kornowski, M. Haase, and H. Weller. Colloidal synthesis and self-assembly of copt$_3$ nanocrystals. *Journal of the American Chemical Society*, 124:11480–11485, 2002.

[125] E.V. Shevchenko, D.V. Talapin, H. Schnablegger, A. Kornowski, O. Festin, P. Svedlindh, M. Haase, and H. Weller. Study of nucleation and growth in the organometallic synthesis of magnitic alloy nanocrystals: The role of nucleation rate in size control of copt$_3$ nanocrystals. *Journal of the American Chemical Society*, 125:9090–9101, 2002.

[126] Long-Pei Shi. Perpendicular magnetic anisotropies in utrathin bcc iron films and surfaces. *J. Phys.: Condens. Matter*, 6:1183–1206, 1994.

[127] O. Sipr. private communication.

[128] G.A. Somorjai. *Introduction to Surface Chemistry and Catalysis*. Wiley, New York, 1994.

[129] L. Soriano, M. Abbate, A. Fernández, A.R. González-Elipe, F. Sirotti, and J.M. Sanz. Oxidation state and size effects in coo nanoparticles. *J.Phys. Chem. B*, 103:6676–6679, 1999.

[130] M. Stampanoni. Magnetic properties of thin epetaxial films investigated by spin-polarized photoemission. *Appl. Physics A*, 49:449–458, 1989.

[131] C. Staudt, R. Heinrich, and A. Wucher. Formation of large clusters during sputtering of silver. *Nuclear Instruments and Methods in Physics Research B*, 164–165:677–686, 2000.

[132] J. Stöhr. *NEXAFS Spectroscopy*. Springer Verlag, 1992.

[133] S. Sun. Recent advances in chemical synthesis, self-assembly, and application of fept nanoparticles. *Advanced Materials*, 18:393–403, 2006.

[134] J. Thiele, C. Boeglin, K. Hricovini, and F. Chevrier. Magnetic circular x-ray-dichroism study of co/pt(111). *Phys. Rev. B*, 53(18):R11934–R11937, May 1996.

[135] B.T. Thole, P. Carra, M. Altarelli, and X. X. Wang. X-ray circular dichroism and local magnetic fields. *Phys. Rev. Lett.*, 70:694, 1993.

[136] B.T. Thole, P. Carra, F. Sette, and G. van der Laan. X-ray circular dicroism as a probe of orbital magnetization. *Phys. Rev. Lett.*, 68(12):1943–1946, 1992.

[137] B.T. Thole and G van der Laan. Branching ratio in x-ray absorption spectroscopy. *Phys. Rev. B*, 38:3158–3171, 1988.

[138] J. Thomassen, F. May, B. Feldmann, M. Wuttig, and H. Ibach. Magnetic live surface layers in fe/cu(100). *Phys. Rev. Lett.*, 69(26):3831–3834, Dec 1992.

[139] A. Thompson, D. Attwood, E. Gullikson, M. Howells, K.-J. Kim, J. Kortright, I. Lindau, P. Pianetta, A. Robinson, J. Scofield, J. Underwood,

D. Vaughan, G. Williams, and H. Winick. *X-RAY DATA BOOKLET*. Lawrence Berkley National Laboratory, 2001.

[140] M.A. Torija, J.P. Pierce, and J. Shen. capping layer induced spin reorientation: Co on fe/cu(100). *Phys. Rev. B*, 63:0924041–0924043, 2001.

[141] F. Tournus, A. Tamion, N. Blanc, A. Hannour, L. Bardotti, B. Prével, P. Ohresser, E. Bonet, T. Epicier, and V. Dupuis. Evidence of ll_0 chemical order in copt nanoclusters: Direct observation and magnetic signature. *Phys. Rev. B*, 77:1444111, 2008.

[142] R. Tsukamoto, M. Muraoka, M. Seki, H. Tabata, and I. Yamashita. Synthesis of copt and $fept_3$ nanowires using the central channel of the tobacco mosaic virus as a biotemplate. *Chem.Mater.*, 19:2389–2391, 2007.

[143] G. van der Laan. Microscopic origin of magnetocrystalline anisotropy in transition metal thin film. *J. Phys.: Condens. Matter*, 10:3239–3253, 1998.

[144] Giovanna Vandoni, Christian Félix, and Carlo Massobrio. Moleculardynamics study of collision, implantation, and fragmentation of $ag7$ on pd(100). *Phys. Rev. B*, 54(3):1553–1556, Jul 1996.

[145] O. Šipr, J. Minár, S. Mankovsky, and H. Ebert. Influence of composition, many-body effects, spin-orbit coupling, and disorder on magnetism of co-pt solid state systems:. *Phys. Rev. B*, 78:1444031–14440312, 2008.

[146] M. R. Weiss, R. Follath, K. J. S. Sawhney, F. Senf, J. Bahrdt, W. Frentrup, A. Gaupp, S. Sasaki, M. Scheer, H. C. Mertins, D. Abramsohn, F. Schäfers, W. Kuch, and W. Mahler. The elliptically polarized undulator beamlines at bessy ii. *Nuclear Instruments and Methods in Physics Research Section A: Accelerators, Spectrometers, Detectors and Associated Equipment*, 467-468(Part 1):449 – 452, 2001.

[147] D. Weller, H. Brändle, G. Gorman, C.-J. Lin, and H. Notarys. Magnetic and magneto-optical properties of cobalt-platinum alloys with perpendicular magnetic anisotropy. *Appl. Phys. Lett.*, 61:2726–2728, 1992.

[148] U. Wiedwald, M. Cerchez, M. Farle, K. Fauth, G. Schutz, K. Zurn, H.-G. Boyen, and P. Ziemann. Effective exchange interaction in a quasi-two-dimensional self-assembled nanoparticle array. *Phys. Rev. B*, 70(21):214412, 2004.

[149] U. Wiedwald, K. Fauth, M. Heßler, H.-G. Boyen, F. Weigl, M. Hilgendorff, M. Giersig, G. Schütz, P. Ziemann, and M. Farle. From colloidal co/coo core/shell nanoparticles to arrays of metallic nanomagnets: Surface modification and magnetic properties. *ChemPhysChem*, 6:2522–2526, 2005.

[150] U. Wiedwald, M. Spasova, E.L. Salabas, M. Ulmenau, M. Farle, Z. Frait, A.F. Rodriguez, D. Arvanitis, N.S. Sobal, M. Hilgendorff, and M. Giersig and. Ratio of orbital-to-spin magnetic moment in co core-shell nanoparticles. *Phys. Rev. B*, 68:0644241–0644245, 2003.

[151] Wikipedia.

[152] R. Wu and A.J. Freeman. Limitation of the magnetic-circular-dichroism spin sum rule for transition metals and importance of the magnetic diploe term. *Phys. Rev. Lett.*, 73:1994–1997, 1994.

[153] R. Wu, D. Wang, and A.J. Freeman. First principles investigation of the validity and range of applicability of the x-ray magnetic circular dichroism sum rule. *Phys. Rev. Lett.*, 71:3581–3584, 1993.

[154] W. Wurth and M. Martins. *The Chemical Physics of Solid Surfaces (chapter12)*. Elsevier, 2007.

[155] M. Wuttig, B. Feldmann, J. Thomassen, F. May, H. Zillgen, A. Brodde, H. Hannemann, and H. Neddermeyer. Structural transformation of fcc iron film on cu(100). *Surface Science*, 291:14–28, 1993.

[156] H. Zillgen, B. Feldmann, and M. Wuttig. Structural and magnetic properties of ultrathin fe films deposited at low temperature on cu(100). *Surface Science*, 321:32–46, 1994.

Acknowledgement

No work can be done in today's research, without the help, good will and cooperation of numerous people. I will not be able to list all that aided this work in one way or another, but I do my best....

First of all I want to thank my PhD supervisors Wilfried Wurth and Michael Martins for giving me the opportunity to explore in this fascinating field of physics, for letting me use the unique experiment ICARUS and for the constant support and advice in the course of this works evolution.

Most of the measurements at the storage ring Bessy II could not have been performed without the help of Michael Wellhöfer, Michael Martins, Sebastian Hankemeier, Christine Boeglin and Paolo Imperia, who all accompanied me on several beam times. My thanks to the Bessy staff that did a fabulous job. In particular I like to mention Helmut Pfau, Kai Godehusen and Gerd Reichardt.

The XMCD measurements of the wetchemically prepared CoPt nanoparticles were done using the high field Magnet of the University Louis Pasteur (ulp) and the Institut de Physique et Chemie des Matériaux de Strasbourg (IPCMS). In this context I gratefully acknowledge Christine Boeglin, who was always there to answer my numerous inquiries and to help hands on to achieve the best possible results for the measurements.

My highest regards for the synthesis and preparation of the wet chemical CoPt nanoparticles and many fruitful discussions belong to Vesna Alesandrovic. In this context I like to thank Professor Horst Weller for enabling and supporting this cooperation project and the technician Andreas Kornovsky for the many SEM/TEM pictures and technical support.

Taking care of ICARUS one has ones hands full, for help and cooperation in the Lab and many inspiring physical discussions I thank Sebastian Hankemeier, Steffen Fiedler and Paolo Imperia. Technical improvements at the chamber could not have been achieved without the help of our group engineer Sven Gieschen, our group technician Holger Meyer, the Uni-HH technical drawing group (Benno Frensche and coworkers), the Uni-HH mechanical workshop (Stephan Fleig and coworkers), the HASYLAB mechanical workshop (Jens Brehling and coworkers), the Uni-HH electronic workshop (Armin Spikofsky and coworkers), the Uni-HH electrician Bernd Schwitzky, the DESY vacuum-welding group (Sebastian Menk and coworkers) and the HASYLAB crystal lab (Manfred Spiwek and Klaudia

Hagemann).

At this place I like to thank the entire work group of Wilfried Wurth for the fun and instructive time I was allowed to enjoy over the years; Alexander Föhlisch for many illumining physical and non physical conversations, my room mates Michael Wellhöfer, Florian Sorgenfrei and Lorenz Jahn for the nice company and the inspiring cooperative physical problem solving hours. Thanks to Stefan Klump for curing me of my POV-Ray ignorance. A special thanks goes to our secretary Marlis Fölck for taking care of so many „little things".

I deeply thank Jens Viefhaus for a lot of support, help and constructive criticism in the final stages of this work.

Last but not least I thank the German ministry for education and research (BMBF) for funding part of this work under the grant BMBF grant 05 KS4 GUB/6, as well as the the DFG that enabled the other part by means of the SFB 668 Projekt A7.

Die VDM Verlagsservicegesellschaft sucht für wissenschaftliche Verlage abgeschlossene und herausragende

Dissertationen, Habilitationen, Diplomarbeiten, Master Theses, Magisterarbeiten usw.

für die kostenlose Publikation als Fachbuch.

Sie verfügen über eine Arbeit, die hohen inhaltlichen und formalen Ansprüchen genügt, und haben Interesse an einer honorarvergüteten Publikation?

Dann senden Sie bitte erste Informationen über sich und Ihre Arbeit per Email an *info@vdm-vsg.de*.

Sie erhalten kurzfristig unser Feedback!

VDM Verlagsservicegesellschaft mbH
Dudweiler Landstr. 99   Telefon +49 681 3720 174
D - 66123 Saarbrücken   Fax   +49 681 3720 1749
www.vdm-vsg.de

Die VDM Verlagsservicegesellschaft mbH vertritt

Printed by Books on Demand GmbH, Norderstedt / Germany